ATOMS AND LIGHT INTERACTIONS

PHYSICS OF ATOMS AND MOLECULES

Recent volumes in the series:

ATOMIC INNER-SHELL PHYSICS
Edited by Bernd Crasemann

ATOMIC PHOTOEFFECT
M. Ya. Amusia

ATOMIC SPECTRA AND COLLISIONS IN EXTERNAL FIELDS
Edited by K. T. Taylor, M. H. Nayfeh, and C. W. Clark

ATOMS AND LIGHT: INTERACTIONS
John N. Dodd

COHERENCE IN ATOMIC COLLISION PHYSICS
Edited by H. J. Beyer, K. Blum, and R. Hippler

COLLISIONS OF ELECTRONS WITH ATOMS AND MOLECULES
G. F. Drukarev

ELECTRON–MOLECULE SCATTERING AND PHOTOIONIZATION
Edited by P. G. Burke and J. B. West

THE HANLE EFFECT AND LEVEL-CROSSING SPECTROSCOPY
Edited by Giovanni Moruzzi and Franco Strumia

ISOTOPE SHIFTS IN ATOMIC SPECTRA
W. H. King

MOLECULAR PROCESSES IN SPACE
Edited by Tsutomu Watanabe, Isao Shimamura, Mikio Shimizu, and Yukikazu Itikawa

PROGRESS IN ATOMIC SPECTROSCOPY, Parts A, B, C, and D
Edited by W. Hanle, H. Kleinpoppen, and H. J. Beyer

QUANTUM MECHANICS VERSUS LOCAL REALISM: The Einstein–Podolsky–Rosen Paradox
Edited by Franco Selleri

RECENT STUDIES IN ATOMIC AND MOLECULAR PROCESSES
Edited by Arthur E. Kingston

THEORY OF MULTIPHOTON PROCESSES
Farhad H. M. Faisal

ZERO-RANGE POTENTIALS AND THEIR APPLICATIONS IN ATOMIC PHYSICS
Yu. N. Demkov and V. N. Ostrovskii

ATOMS AND LIGHT INTERACTIONS

JOHN N. DODD
University of Otago
Otago, New Zealand

PLENUM PRESS • NEW YORK AND LONDON

Library of Congress Cataloging-in-Publication Data

Dodd, John N.
Atoms and light : interactions / John N. Dodd.
p. cm. -- (Physics of atoms and molecules)
Includes bibliographical references and index.
ISBN 0-306-43741-4
1. Electromagnetic radiation. 2. Electromagnetic interactions.
3. Light. 4. Atoms. I. Title. II. Series.
QC475.D63 1991
539.2--dc20 91-9028
CIP

ISBN 0-306-43741-4

A Division of Plenum Publishing Corporation
233 Spring Street, New York, N.Y. 10013

Printed in the United States of America

PREFACE

This book discusses the interaction of light with atoms, concentrating on the semiclassical descriptions of the processes. It begins by discussing the classical theory of electromagnetic radiation and its interaction with a classical charged dipole oscillator. Then, in a pivotal chapter, the interaction with a free charge is described (the Compton effect); it is shown that, in order to give agreement with observation, certain quantum rules must be introduced. The book then proceeds to discuss the interaction from this point of view—light always being described classically, atoms described quantum-mechanically, with quantum rules for the interaction. Subsequent chapters deal with stimulated emission and absorption, spontaneous emission and decay, the general problem of light stimulating and being scattered from the two-state atom, the photoelectric effect, and photoelectric counting statistics. Finally the author gives a personal view on the nature of light and his own way of looking at certain paradoxes.

The writing of this book was originally conceived as a collaboration between the present author and a colleague of former years, Alan V. Durrant. Indeed, some preliminary exchange of ideas took place in the mid-1970s. But the problems of joint-authorship from antipodean positions proved too difficult and the project was abandoned. I would like to record my indebtedness to him for the stimulation of this early association.

I also acknowledge the encouragement of my colleagues at the University of Otago. Special reference must be made to D. M. Warrington and A. E. Musgrave for reading and commenting on certain parts of the text, to R. L. Dowden for his help with word-processing, to R. J. Ballagh for his assistance with graphics and the discussion of the physics that lay behind them, and finally to W. J. Sandle, a most stimulating colleague.

J. N. Dodd
Dunedin

ACKNOWLEDGMENT

I want to acknowledge a long association with George W. Series since our collaboration in his laboratory in Oxford over thirty years ago on the modulation of light.* Following the excitement of this work, I have returned from time to time (perhaps too often) to the semiclassical problem. I thank him for his stimulating friendship.

* J. N. Dodd and G. W. Series, Theory of Modulation of Light in a Double Resonance Experiment, *Proc. R. Soc. London Ser. A*, **263**, 353–370 (1961).

Acknowledgments

I want to acknowledge a long-standing association with Dr. G. W. Series since our collaboration in the laboratory in Oxford thirty years ago on the modulation of light following the excitement of the atoms.[*] I have from time to time (perhaps too often) in the semiclassical [illegible] Chant [illegible] him for his stimulating friendship.

* J. N. Dodd and G. W. Series, Theory of Modulation of Light in a Double Resonance Experiment, Proc. R. Soc. London Ser. A 263, 353–370 (1961).

Contents

CHAPTER 1
INTRODUCTION AND HISTORY
1.1. The Nature of Light 1
1.2. Electromagnetic Theory 2
1.3. The Interaction Process and its Quantum Nature 5
1.4. This Book 5
Suggested Readings 6
References 6

CHAPTER 2
CLASSICAL RADIATION
2.1. The Electromagnetic Field of an Accelerating Charge . . . 7
2.2. The Radiation of Energy from an Accelerating Charge . . . 11

CHAPTER 3
THE OSCILLATING CHARGE
3.1. The Equation of Motion of an Oscillator 13
3.2. Specification of the Polarization 17
3.3. Polychromatic Oscillation 19
3.4. The Equation of Motion of an Oscillating Charge 21
3.5. Radiation from a Freely Oscillating Charge 25
3.6. Radiation from a Driven Oscillating Charge 26

CHAPTER 4
SCATTERING OF RADIATION FROM A CHARGE DRIVEN BY AN ELECTROMAGNETIC FIELD
4.1. The Case of the Monochromatic Field 27
4.2. The Case of Broadband Radiation 31

CHAPTER 5
INTENSITY, ENERGY DENSITY, THE POYNTING VECTOR, AND THEIR SPECTRAL DISTRIBUTIONS
5.1. The Intensity of the Radiation Field 35
5.2. Energy Density and the Poynting Vector 36
5.3. Spectral Distributions 41

CHAPTER 6
THE INTERACTION OF A BEAM OF ELECTROMAGNETIC RADIATION WITH A FREE ELECTRIC CHARGE—THE COMPTON EFFECT
6.1. The Compton Effect 45
6.2. A Classical Theory of the Compton Effect 46
6.3. The Laws of Interaction Between Radiation and Matter . . 50
6.4. The Scattered State 56

CHAPTER 7
THE QUANTUM STRUCTURE OF THE ATOM
7.1. Transitions in Atoms 57
7.2. Allowed States 59
7.3. State Vectors 62
7.4. Labeling the Allowed State (Eigenstate) Vectors 65
7.5. The Configuration 67
7.6. The Matrix Elements of an Operator 73
7.7. Selection Rules for Electric Dipole Radiative Transitions . 76
7.8. Superposition States and the Equation of Motion 79
7.9. The Correspondence Principle 83
7.10. The Wave Function 84
References 86

CHAPTER 8
THE EINSTEIN *A* AND *B* COEFFICIENTS
8.1. Populations and Transition Rates 89
8.2. The Classical Theory for the *A* Coefficient 91
8.3. The Radiative Lifetime and Decay Constant 92
8.4. The Classical Theory for the *B* Coefficient 95

CHAPTER 9
THE SEMICLASSICAL TREATMENT OF STIMULATED ABSORPTION AND EMISSION
9.1. The Quantum Equation of Motion 97
9.2. Stimulation by Monochromatic Radiation 99

9.3. Stimulation by Polychromatic Radiation 101
9.4. Deduction of the Einstein B Coefficient 102
Reference 104

CHAPTER 10
THE SEMICLASSICAL DESCRIPTION OF SPONTANEOUS DECAY
10.1. The One-Way Nature of Decay 105
10.2. Spontaneous Emission as Derived from Radiation Reaction 107
10.3. Spontaneous Emission as Derived from the Zero-Point Vacuum Field 110
10.4. Remarks on Semiclassical Treatments for Spontaneous Decay 117
10.5. The Matrix Elements of $\hat{H}_D$ 120
References 120

CHAPTER 11
THE GENERAL OPTICAL TRANSITION
11.1. The Two-Level Atom 121
11.2. Stimulated Transitions at the Rabi Frequency 122
11.3. The Two-Level Atom with Decay 126
11.4. The Steady-State Solution 131
11.5. Radiation from a Driven Atom 136
References 140

CHAPTER 12
THE PHOTOELECTRIC EFFECT
12.1. A General Discussion of the Photoelectric Effect 141
12.2. The Photoelectric Differential Cross Section 145
12.3. The Photoelectric Differential Cross Section—An Alternative View 149
12.4. The Photoelectric Differential Cross Section—A More General Result 153
12.5. The Photoelectric Cross Section for the Hydrogen Atom . . 156
12.6. Remarks 159
Reference 160

CHAPTER 13
OPTICAL COHERENCE AND COUNTING STATISTICS
13.1. Interference and the Coherence of Light 161
13.2. The First-Order Coherence Function 163
13.3. The Second-Order Coherence Function 164
13.4. Partially Coherent Light 167

13.5. Photoelectric Statistics 174
13.6. Nonclassical Light 176
References 180

CHAPTER 14
WHAT *IS* LIGHT?
14.1. Models of Light 181
14.2. Emission and Detection 184
14.3. An Alternative Description—The Luminal Frame 188
14.4. Source–Detector Interaction 189
14.5. Cascade Optical Transitions 191
14.6. The Same Experiment with Particles 199
14.7. The Answer 200
References 201

APPENDIX 1
TIME AVERAGING 203
A1.1. Truncated Time Averaging 204
A1.2. Exponential Averaging 205
A1.3. Averaging over a Finite Time Interval 206
A1.4. The Energy Density and Spectral Distribution of "Broad-Band" or "White" Light 206
A1.5. The Monochromatic Field 207
A1.6. The Modulated Field Arising from the Coherent Superposition of Two Frequencies 208

APPENDIX 2
ENSEMBLE AVERAGING 211

APPENDIX 3
THE ZERO-POINT VACUUM FIELD 217
References 222

APPENDIX 4
AN INVARIANT FORM FOR ANGULAR MOMENTUM—JUSTIFICATION FOR EQ. (6.17) 223

APPENDIX 5
FUNCTIONS 225
A5.1. The Dirac Delta Function 225
A5.2. The Exponential and the Lorentz Functions 228

A5.3. The Square Pulse and the Fraunhofer Function 232
A5.4. The Peaked Functions 234

APPENDIX 6
WAVE FUNCTIONS AND BRA-KETS 235

APPENDIX 7
THE DENSITY OF WAVES IN A BOX 241
A7.1. The Density of Electromagnetic Field Modes 242
A7.2. The Density of Free Particle States 243

INDEX 245

CHAPTER 1

INTRODUCTION AND HISTORY

1.1. THE NATURE OF LIGHT

The Greeks, of course, had a word for it: $\phi\omega\sigma$, light. During the first century B.C., the Roman poet Lucretius wrote a great poem "On the Nature of the Universe."[(1)] (One may marvel that the poet of those days was interested in and understood so much science. It is perhaps more likely that Lucretius, the scientist, used poetry to express the intellectual wonderment of his subject. The tragedy is that the scientist of today is unable to express the wonder of his story in poetical ways.) His description is based on the Greek view of nature as expressed by Epicurus. Light is described in two ways: as an emission of "atoms" from a luminous source such as the sun, and also as a sloughing off of a very thin outer shell of an object, which conveys to our senses the shape, texture, color, and smell of the object.

Greek science was based on philosophical and logical deductions based on the observations and sensations of man. These sensations were the truth about the nature of things. But, remarkable as many of these statements were, there was no routine testing of the hypotheses. That had to await the flowering of experimental science. The most important next steps were made by the greats of the sixteenth and seventeenth centuries—Galileo, Huygens, and Newton.

Huygens proposed that light was a wave motion and put forward the construction, that goes by his name, for using secondary spherical wavelets for establishing the successive positions of the wave front and hence the path of the light. By this means he successfully explained the observed laws of reflection and refraction of light. But to the question "waves of what, in what?" he gave no answer.

Newton was perhaps a bit closer to the views held by the Greeks in that he proposed (not dogmatically) a corpuscular theory. With this he could also explain the laws of reflection and refraction but only, for the latter, if the corpuscles of light traveled faster in transparent material than in air (or vacuum).

To the majority of scientists the wave theory of light became vindicated by the optical interference experiments of Young in 1801, and their quantitative description. Dogma was not dead, however, and in certain quarters Young's views aroused protest, derision, and abuse. Brougham, later to become Lord Chancellor of England, in reviewing Young's paper for the Edinburgh Review, wrote:

> We wish to raise our feeble voice against innovations that can have no other effect than to check the progress of science and renew all those wild phantoms of the imagination which Bacon and Newton put to flight from her temple. We wish to recall philosophers to the strict and severe methods of investigation.[2]

There speaks the voice of assumed authority!

However, the wave theory was to flourish, largely because of the theoretical and experimental studies of the three great French optical physicists of the nineteenth century, Fresnel, Foucault, and Fizeau. They rediscovered optical interference, using mirrors and prisms to superimpose beams of light. They measured the velocity of propagation and showed that it was slower in water than in air (vacuum). They discovered and successfully explained the phenomena of diffraction and polarization. They placed the wave theory of light on a proper mathematical basis. Light propagates through vacuum with a velocity close to 3×10^8 m s^{-1} and with wavelengths in the region of 400–700 nm for visible light. Furthermore, the measured velocity in water is a factor of 1.33 less than in air (vacuum), exactly the value required to explain the phenomenon of refraction based on Huygen's construction. The construction itself was placed on a proper mathematical footing by the wave theory of Fresnel.

But still there was no answer to the question "waves of what, in what?" To say that light is undulations in the *luminiferous aether* really only invents a couple of new words. And if this aether, which has to pervade everything through which light can pass including vacuum, has mechanical properties required to transmit transverse waves, it would have to be pretty powerful stuff.

1.2. ELECTROMAGNETIC THEORY

The concept of electric and magnetic vectors had grown up during the nineteenth century to explain the *action at a distance* between two electrical sources of charge and current. The electric vector **E** and the magnetic vector **B** are defined through the equation

$$\mathbf{F} = q(\mathbf{E} + \mathbf{v} \times \mathbf{B}) \tag{1.1}$$

where **F** is the force acting on the test charge q due to its interaction with a source. To establish all the components of **E** and **B**, the force would have to be measured on q under a chosen variety of conditions, e.g., with q at rest when **E** is established and then with certain chosen velocities to find **B**. This equation, called the Lorentz force equation, can be put in terms of densities:

$$\mathbf{f} = \rho\mathbf{E} + \mathbf{J} \times \mathbf{B} \tag{1.2}$$

where **f** is the force per unit volume on an element of space where the local charge density is ρ and the local current density is **J**.

The fields **E** and **B** are created by source charges Q moving with various velocities **v** to create a distribution of charge density ρ_s and current density $\mathbf{J}_s$ throughout the source region. The values of **E** and **B** at any point in space can be deduced from equations like Coulomb's law and the Biot–Savart law which relate the fields to their sources.

Clerk Maxwell, in 1862–1864, established a set of differential equations governing the vectors **E** and **B** at any point in space-time in relation to the charge and current densities, ρ and **J**, *at the same space-time point.* These equations,

$$\nabla \times \mathbf{B} - \frac{1}{c^2}\frac{\partial \mathbf{E}}{\partial t} = \mu_0 \mathbf{J} \qquad \text{(Ampère's law)} \tag{1.3}$$

$$\nabla \cdot \mathbf{E} = \mu_0 c^2 \rho \qquad \text{(Gauss's law for } \mathbf{E}\text{)} \tag{1.4}$$

$$\nabla \times \mathbf{E} + \frac{\partial \mathbf{B}}{\partial t} = 0 \qquad \text{(Faraday's law)} \tag{1.5}$$

$$\nabla \cdot \mathbf{B} = 0 \qquad \text{(Gauss's law for } \mathbf{B}\text{)} \tag{1.6}$$

together with Eq. (1.2) form a complete statement about electromagnetic interaction between charges and currents. In Eq. (1.4), ρ is the *total* charge density at the space-time point and will contain, in general, a contribution from *bound* charges deriving from the polarization of material as well as from *free* charges:

$$\rho = \rho_{\text{free}} - \nabla \cdot \mathbf{P}$$

In Eq. (1.3), **J** is the *total* current density and will contain, in general, contributions from the *bound* currents deriving from changing polarization and from the magnetization of the material as well as from *free* currents:

$$\mathbf{J} = \mathbf{J}_{\text{free}} + \frac{\partial \mathbf{P}}{\partial t} + \nabla \times \mathbf{M}$$

A static solution for Maxwell's equations, i.e., when ρ is constant in time and **J** is zero over a small volume, yields the Coulomb equation for the electric field **E** of a source charge. When the source charges are in steady motion, creating a steady current, the solution involves a magnetic field **B** also, given by the Biot-Savart equation. But we shall need more general solutions for moving and *accelerating* point sources, which are quoted without proof in Chapter 2.

In establishing Maxwell's equations two physical constants have emerged, μ_0 and c. The permeability of vacuum, μ_0, establishes a proportionality between **B** and **J** and hence for the force between current elements. It is therefore used to define an appropriate unit of current. In the S.I. system it has the value $\mu_0 = 4\pi \times 10^{-7}$ N A^{-2} exactly. From the definition of the ampere, using this value for the proportionality, the unit of charge, the coulomb, is established. The value of the constant c in the combination $\mu_0 c^2$ is then determined by measuring the interactive force between two charges measured in units of coulombs. It turns out that c has a value close to 3×10^8 m s^{-1}.

Maxwell demonstrated theoretically that an oscillating source will generate oscillating fields that propagate through space with the velocity c. He remarked that this

> agrees so exactly with velocity of light calculated from the optical experiments of M. Fizeau, that we can scarcely avoid the inference that light consists of transverse undulations of the same medium which is the cause of electric and magnetic phenomena.[3]

The velocity c turns out to be a very special kind of velocity—it is a *physical constant*, which is a rather peculiar thing for a velocity to be. It was this, together with the fact that Maxwell's equations transform between two observers in a different way than did the then accepted laws of mechanics, and the fact that, according to experiment, it appeared that the velocity of light was indeed independent of the motion of the source or the observer, that led Einstein (1905) to propound the *theory of relativity* (or *invariance* as it would more properly be called). This led to the first great revolution in twentieth century physics.

Light and other forms of radiation had indeed become accepted as electromagnetic waves. Subsequently the measured velocity of light in vacuum was measured to such high accuracy that, in 1983, it was deemed appropriate to define it exactly:

$$c = 299\,792\,458 \text{ m s}^{-1} \qquad \text{(exactly)}$$

This is tantamount to defining the unit of length (meter) in terms of a specified fraction of the distance that light travels in the unit of time (second).

In this way the constants μ_0 and c appearing in Maxwell's equations have exactly defined values.

1.3. THE INTERACTION PROCESS AND ITS QUANTUM NATURE

In spite of the success that had been achieved by this electromagnetic wave theory for light, all was not well. The electromagnetic wave transports energy away from the source, and this energy would be distributed over all regions of the wave. Yet it is observed that, when this wave interacts with charged matter, such as setting a free electron into motion, exciting or ionizing an atom, the energy interchanged is highly localized. This led to new theories about the interaction process, which gave rise in the early twentieth century—Planck (1900), Einstein (1905), and Bohr (1913)—to the *quantum theory*, the second great revolution of twentieth century physics. In this interactive process the radiation is regarded as corpuscular in nature giving the localization necessary to explain the observation that an atom absorbs a specified amount of energy from the field. It seems that light has two different natures as described by the two different models—it propagates as a wave, it interacts like a particle.

1.4. THIS BOOK

We set out to describe the interaction between light (radiation) and atoms (matter). In Chapters 2–5 the classical description of the electromagnetic field is given, how it relates to its sources, and how it interacts with and scatters from charged particles.

In Chapter 6 we consider the experimental facts of the simplest interaction—radiation with a charged particle—and are led to the conclusion that some quantum rules are required to govern the changes that take place. However, it is our point of view that, at this point at least, it is unnecessary to abandon the wave theory; the rules apply to the interaction process because of the quantum nature of the behavior of a particle system.

Chapter 7 gives a brief description of the quantum structure of the atom, sufficient for our later needs.

The later chapters then proceed to discuss the interaction of electromagnetic radiation (a wave phenomenon) with atoms (a quantum structure)—transitions, absorption and decay, the photoelectric effect, coherence, and counting statistics. This way of looking at things is often referred to as the *semiclassical viewpoint*. It may well be that the semiclassical viewpoint falls down at some stage and is unable to predict correctly certain phenomena;

my own view is that it succeeds much more widely than it is given credit for. Consequently this book proceeds along the semiclassical path. Even if it is not justified from the point of view of many physicists, it is still useful for another reason. Even if the quantum nature of radiation (quantum electrodynamics or QED as it is often called) is required, the underlying physics needs a firm understanding of its classical basis.

Finally, in Chapter 14, I present a personal view on the nature of light. Now, read on.

SUGGESTED READINGS

This book contains no extensive bibliography of the field; useful references are given from time to time in the text. Overall I would like to make reference to a few sources only:

E. U. CONDON AND G. H. SHORTLEY, *The Theory of Atomic Spectra* (University Press, Cambridge, 1935; reprint 1953).

E. WHITTAKER, *A History of the Theories of Aether and Electricity*, Vols. 1 and 2 (Thomas Nelson and Sons, London, 1951).

M. BORN AND E. WOLF, *Principles of Optics* (Pergamon Press, London, 1959).

C. COHEN-TANNOUDJI, B. DIU, AND F. LALOË, *Quantum Mechanics*, Vols. 1 and 2 (John Wiley and Sons, New York, 1977).

G. W. SERIES, A Semi-classical Approach to Radiation Problems, *Physics Reports* **43**(1), (1978).

R. LOUDON, *The Quantum Theory of Light*, 2nd ed. (Clarendon Press, Oxford, 1983).

REFERENCES

1. TITUS LUCRETIUS CARUS, *On the Nature of the Universe*, Translated by R. E. Latham (Penguin Books, Harmondsworth, Middlesex, 1951).
2. H. P. BROUGHAM, quoted in F. K. Richtmeyer, E. H. Kennard, and J. N. Cooper, *Introduction to Modern Physics* (McGraw-Hill, New York, 1955), p. 29.
3. J. C. MAXWELL, quoted in F. K. Richtmeyer, E. H. Kennard, and J. N. Cooper, *Introduction to Modern Physics* (McGraw-Hill, New York, 1955), p. 41.

CHAPTER 2

CLASSICAL RADIATION

2.1. THE ELECTROMAGNETIC FIELD OF AN ACCELERATING CHARGE

We present here a brief summary of the principal results from classical radiation theory which will be needed later.

We begin by specifying the field vectors (**E** and **B**) at field point P, at position **x** at time t, created by a particle of charge Q at the position $\mathbf{x}'$ moving with velocity $\mathbf{v}'$ and acceleration $\mathbf{a}'$, both in general being functions of time. The geometry is illustrated in Fig. 1, which is drawn for the case of the charged particle moving around a circle at 1/5 of the velocity of light ($Q'P = 5Q'Q = 5Q'Q^*$). The *present* position (i.e., at time t) of the charge at some point along its path is Q. However, owing to the fact that a field propagates in space at velocity c, it is the position Q' with coordinate $\mathbf{x}'$, the velocity $\mathbf{v}'$, and the acceleration $\mathbf{a}'$ at some earlier (retarded) time t' that determines the field at P at time t. If Q' is the *retarded* position of the source point, then we have

$$\begin{aligned} Q'P &= R \\ &= [(x-x')^2+(y-y')^2+(z-z')^2]^{1/2} \\ &= c(t-t') \end{aligned} \tag{2.1}$$

Besides the present position Q and the retarded position Q', the "virtual" position Q^* is also shown, i.e., the position the charge would have reached at time t if it had continued to move with constant velocity $\mathbf{v}'$ from the retarded position Q'; thus $Q' \to Q^* = \mathbf{v}'(t-t') = R(\mathbf{v}'/c)$.

The electric field vector $\mathbf{E}(\mathbf{x}, t)$ at the field point P at time t is given, according to well-known electromagnetic theory, by

$$\mathbf{E}(\mathbf{x}, t) = \frac{\mu_0}{4\pi} Q \left[\frac{c^2(\mathbf{R} - R\mathbf{v}'/c)}{\gamma^2 s^3} + \frac{\mathbf{R} \times ((\mathbf{R} - R\mathbf{v}'/c) \times \mathbf{a}')}{s^3} \right] \tag{2.2}$$

in which $\gamma^2 = 1/(1 - v'^2/c^2)$ and $s = R - \mathbf{R} \cdot \mathbf{v}'/c$. We have replaced the more usual ε_0 by $1/\mu_0 c^2$. Every quantity within the square brackets is to be evaluated at the retarded time $t' = t - R/c$; to emphasize this, the velocity and acceleration of the charge have been written as primed quantities. In terms of the diagram Fig. 1, $s = MP$, and $\mathbf{R} - R\mathbf{v}'/c = Q^* \to P$.

In the nonrelativistic limit ($v' \to 0$, $\gamma \to 1$, $s \to R$), the first term in Eq. (2.2) gives the Coulomb formula. In its relativistically corrected form, the electric field expressed by this term points *radially away from the virtual position of the charge*, Q^*, i.e., in the direction $Q^* \to P$. This Coulomb field falls off with distance as $1/R^2$; it is shown as E_C in Fig. 1.

The second term of Eq. (2.2) depends on the acceleration of the charge. This contribution to the electric field, which is orthogonal to $\mathbf{R}$, falls off with distance only as $1/R$; it is shown as $\mathbf{E}_r$ in Fig. 1 and is referred to as the *radiation field*. The electric energy density calculated from this term falls off as $1/R^2$ and contributes to a radiation of energy.

The magnetic field vector at the point $\mathbf{x}$ at time t is

$$\mathbf{B}(\mathbf{x}, t) = \frac{\mathbf{R} \times \mathbf{E}}{Rc}$$

$$= \frac{\mu_0}{4\pi} Q \left[\frac{\mathbf{v}' \times \mathbf{R}}{\gamma^2 s^3} + \frac{\mathbf{R} \times (\mathbf{R} \times ((\mathbf{R} - R\mathbf{v}'/c) \times \mathbf{a}'))}{Rcs^3} \right] \tag{2.3}$$

The first term in Eq. (2.3) gives the Biot–Savart formula corrected for relativistic effects. The second term depends on the acceleration $\mathbf{a}'$ and is proportional to $1/R$; it is the *radiation field* $\mathbf{B}_r$. The magnetic energy density calculated from this term falls off as $1/R^2$ and contributes to a radiation of energy.

It is to be noted further that both of these final terms, the radiation fields $\mathbf{E}_r(\mathbf{x}, t)$ and $\mathbf{B}_r(\mathbf{x}, t)$, are transverse to the radius vector $\mathbf{R}$ from the *retarded* position of the source, as well as being orthogonal to each other. The Poynting vector derived from these represents a radiation of energy in the direction of $\mathbf{R}$.

In this text the interest is in the radiation fields; consequently we shall consider only the final terms in Eqs. (2.2) and (2.3). Furthermore, the interest is primarily in the nonrelativistic situation; consequently we shall mainly use nonrelativistic approximations. Finally we shall be concerned with sources of radiation of atomic dimensions; it is then appropriate to choose an origin of coordinates at, or close to, the center of cyclic oscillation in

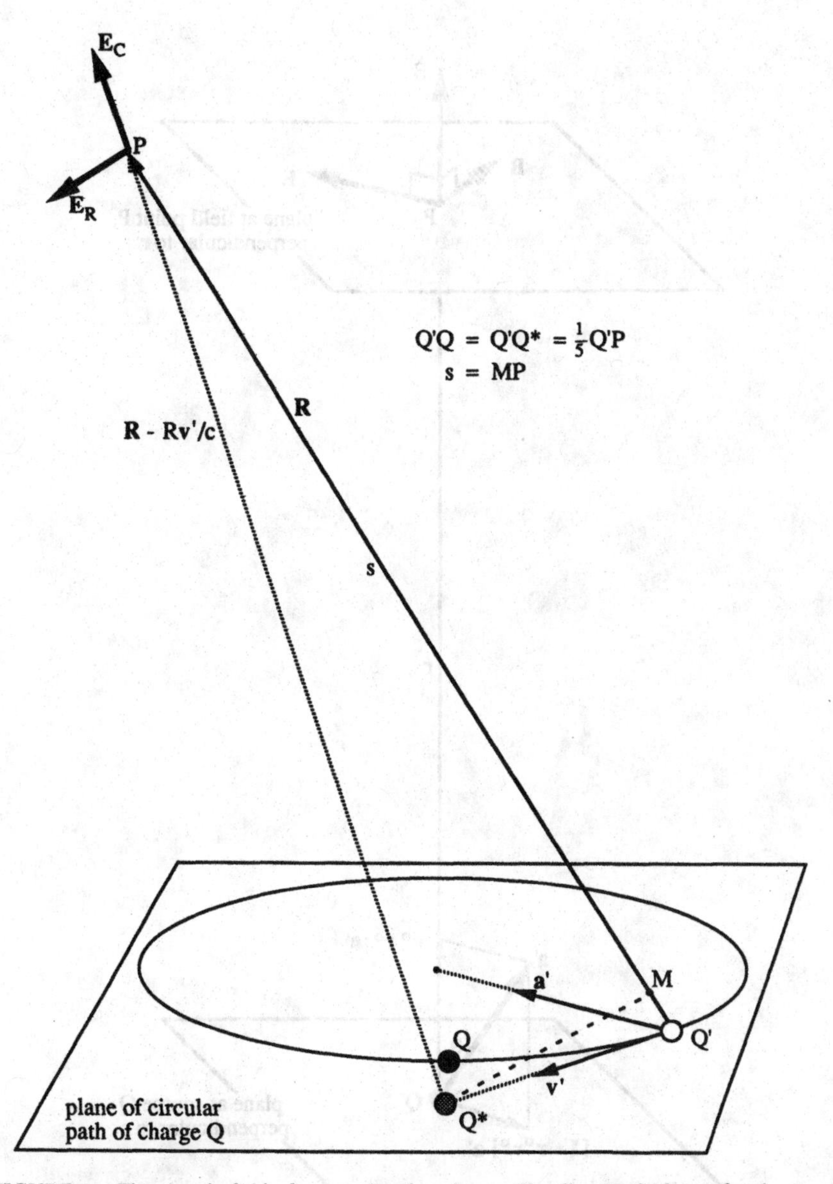

FIGURE 1. The electric field of an accelerating charge. The diagram is drawn for the case of a charge Q moving around a circular path at about one-fifth of the speed of light. $\mathbf{E}_C$, the Coulomb field, is directed along $Q^* \rightarrow P$. $\mathbf{E}_R$, the radiation field is perpendicular to $Q' \rightarrow P$, and lies in the same plane as $\mathbf{R}$ and $\mathbf{a}'$.

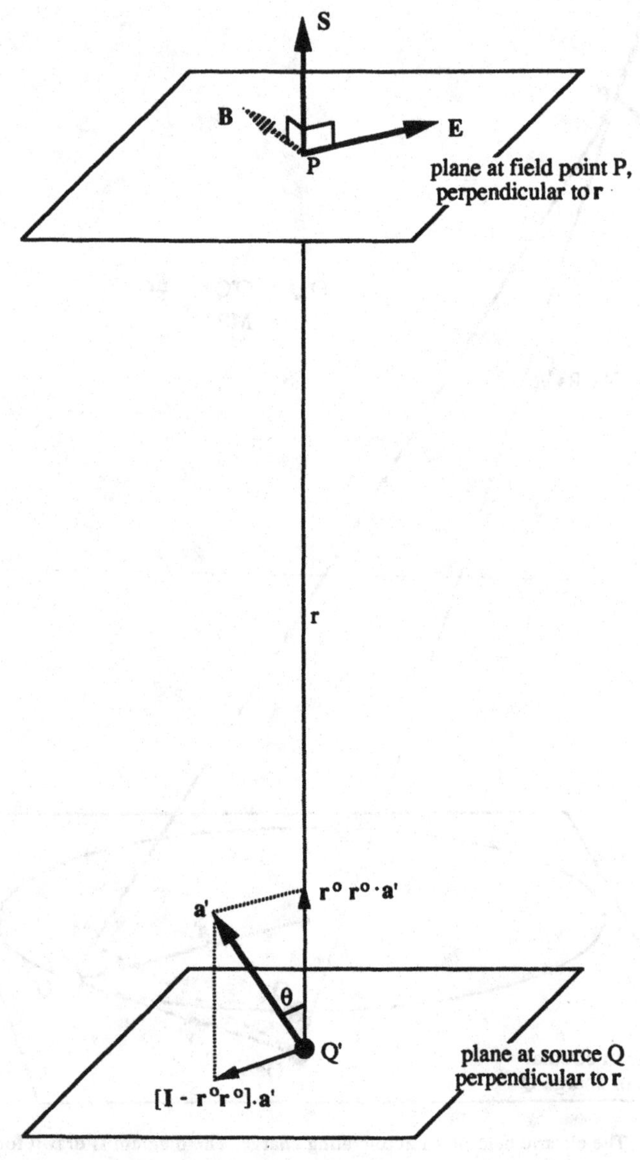

FIGURE 2. The relationships of **E**, **B**, and **S** at the field point P to the acceleration **a**′ of the charge at the retarded position Q'.

the source, and replace **R** by **r**, the radius vector of the field point from an origin in the small source. The radiation field can now be specified by

$$\mathbf{E}(\mathbf{r}, t) = \frac{\mu_0}{4\pi} Q \frac{\mathbf{r}^0 \times (\mathbf{r}^0 \times \mathbf{a}')}{r} \tag{2.4}$$

$$\mathbf{B}(\mathbf{r}, t) = \frac{\mu_0}{4\pi} Q \frac{\mathbf{r}^0 \times (\mathbf{r}^0 \times (\mathbf{r}^0 \times \mathbf{a}'))}{cr} \tag{2.5}$$

where the unit vector $\mathbf{r}^0 = \mathbf{r}/r$. The effect of retardation is now expressed entirely by the prime on the acceleration, i.e., $\mathbf{a}'$ is the acceleration of the charge at time t'. It is convenient to express these in a slightly different form:

$$\mathbf{E}(\mathbf{r}, t) = -\frac{\mu_0}{4\pi} \frac{Q}{r} [\mathbf{I} - \mathbf{r}^0\mathbf{r}^0] \cdot \mathbf{a}' \tag{2.6}$$

$$\mathbf{B}(\mathbf{r}, t) = -\frac{\mu_0}{4\pi} \frac{Q}{cr} \mathbf{r}^0 \times \mathbf{a}' \tag{2.7}$$

where **I** is the idem dyadic tensor. In the present geometry, at the point $P(r, \theta, \phi)$, $\mathbf{I} = \mathbf{r}^0\mathbf{r}^0 + \boldsymbol{\theta}^0\boldsymbol{\theta}^0 + \boldsymbol{\phi}^0\boldsymbol{\phi}^0$, where $\mathbf{r}^0$, $\boldsymbol{\theta}^0$, $\boldsymbol{\phi}^0$ are the unit vectors pointing in the directions in which P would move if r, θ, and ϕ alone were to increase, respectively. (The notations of spherical geometry are shown in Fig. 3, page 19.) The tensor operation $[\mathbf{I} - \mathbf{r}^0\mathbf{r}^0]\cdot$ projects $\mathbf{a}'$ onto a plane perpendicular to $\mathbf{r}^0$ as shown in Fig. 2.

2.2. THE RADIATION OF ENERGY FROM AN ACCELERATING CHARGE

The Poynting vector at the point P for the radiation fields is

$$\mathbf{S}(\mathbf{r}, t) = \frac{\mathbf{E} \times \mathbf{B}}{\mu_0}$$

$$= \frac{\mu_0}{4\pi} \frac{Q^2}{4\pi c r^2} [\mathbf{a}' \cdot \mathbf{a}' - (\mathbf{a}' \cdot \mathbf{r}^0)^2] \mathbf{r}^0$$

$$= \frac{\mu_0}{4\pi} \frac{Q^2}{4\pi c r^2} \mathbf{a}' \cdot \mathbf{a}' \sin^2\theta \, \mathbf{r}^0 \tag{2.8}$$

where θ is the angle between $\mathbf{a}'$ and $\mathbf{r}^0$ (see Fig. 2). By integrating this over a sphere we obtain Larmor's formula for the radiated power:

$$P_{\text{RAD}}(t') = \frac{\mu_0}{4\pi} \frac{2Q^2}{3c} \mathbf{a}' \cdot \mathbf{a}' \tag{2.9}$$

The result has been deduced for the case $\mathbf{v}' = 0$, i.e., it gives the rate of radiation from the accelerating charge in its own, instantaneous, rest frame. Since power is the rate of change of energy, $P = dW/dt$. W and t are each fourth components of Lorentz 4-vectors, $(\mathbf{p}, W/c)$ and $(\mathbf{x}, ct)$, respectively, and therefore transform relativistically in exactly the same manner; P is therefore Lorentz invariant. Indeed, if the power is deduced more generally from the final terms of Eqs. (2.2) and (2.3), the following results are obtained:

(i) if $\mathbf{v}'$ and $\mathbf{a}'$ are collinear: $$P_{\text{RAD}} = \frac{\mu_0}{4\pi}\frac{2Q^2}{3c}(\gamma^3 a')^2$$

In this case ($\mathbf{a}'$ parallel to the transformation velocity), special relativity shows that $\gamma^3 a'$ is invariant.

(ii) if $\mathbf{v}'$ and $\mathbf{a}'$ are orthogonal: $$P_{\text{RAD}} = \frac{\mu_0}{4\pi}\frac{2Q^2}{3c}(\gamma^2 a')^2$$

In this case ($\mathbf{a}'$ perpendicular to the transformation velocity), $\gamma^2 a'$ is invariant. Thus Eq. (2.9) represents the radiant power from any accelerating charge in terms of the acceleration *measured in the instantaneous proper frame of the moving charge.*

Equation (2.9) does not, of course, represent the total instantaneous rate of loss of energy from a moving charge; the earlier terms in Eqs. (2.2) and (2.3) also give a contribution. However, such energy, derived from the *local fields*, is not radiated but is exchanged between the mechanical energy of the sources and the stored energy of the local field. It does not provide any net damping to the motion of the charge.

CHAPTER 3

THE OSCILLATING CHARGE

3.1. THE EQUATION OF MOTION OF AN OSCILLATOR

The equation of motion of a mass m is

$$m\ddot{\mathbf{x}} = \mathbf{F}_{\text{applied}} + \mathbf{F}_{\text{restoring}} + \mathbf{F}_{\text{dissipation}} \tag{3.1}$$

where the force acting on the mass has been broken into three contributions appropriate to an oscillating system. We shall consider that the system moves under a linear elastic restoring force,

$$\mathbf{F}_{\text{restoring}} = -m4\pi^2\nu_0^2\mathbf{x}$$

such that, under free undamped oscillation, the system would vibrate with a natural frequency ν_0. The dissipation force is usually taken to be proportional to the velocity, $\mathbf{F}_{\text{dissipation}} = -m\gamma\dot{\mathbf{x}}$, where γ is the damping constant. These choices lead to the usual equation of motion for a damped harmonic oscillator:

$$\ddot{\mathbf{x}} + \gamma\dot{\mathbf{x}} + 4\pi^2\nu_0^2\mathbf{x} = \frac{\mathbf{F}(t)}{m} \tag{3.2}$$

where the applied force, in general a function of time, is now written as $\mathbf{F}(t)$. The solutions of this equation are well known and are quoted here for reference:

i. Free Oscillation, $\mathbf{F}(t) = 0$. The solution is

$$x = \sqrt{2}A_0\, e^{-\gamma t/2}\cos(2\pi\nu' t + \alpha) \tag{3.3}$$

The free frequency is

$$\nu' = \nu_0\left(1 - \frac{\gamma^2}{16\pi^2\nu_0^2}\right)^{1/2} \tag{3.4}$$

A_0 and α are the rms amplitude and the phase, respectively, at $t = 0$, and are determined from the initial conditions.

ii. Steady-State Forced Oscillation. The oscillatory applied force is

$$\mathbf{F}(t) = q\mathbf{E}(t) = \boldsymbol{\pi}^0\sqrt{2}\,qE_0\cos(2\pi\nu_L t)$$

supplied by an oscillating electric field of *root-mean-square* amplitude E_0 and frequency ν_L (the subscript L indicates a unique frequency of a monochromatic field supplied, for example, by a laser) acting on a particle of charge q and mass m. In place of the *real* physical quantities $\mathbf{x}(t)$, $\mathbf{E}(t)$, and $\mathbf{F}(t)$ above it is convenient to introduce complex notation, and, in fact, a more compact method of writing general states of polarization is thereby achieved (discussed in Section 3.2 below). The applied electric field and force are

$$\mathbf{E}(t) = \mathbf{E}^+(t) + \mathbf{E}^-(t) = \frac{E_0}{\sqrt{2}}(\boldsymbol{\pi}^+\, e^{-i2\pi\nu_L t} + \boldsymbol{\pi}^-\, e^{+i2\pi\nu_L t}) \tag{3.5}$$

where

$$\mathbf{E}^+(t) = \boldsymbol{\pi}^+\frac{E_0}{\sqrt{2}}\, e^{-i2\pi\nu_L t}$$

$$\mathbf{F}(t) = \mathbf{F}^+(t) + \mathbf{F}^-(t), \qquad \text{where } \mathbf{F}^+(t) = \boldsymbol{\pi}^+\frac{qE_0}{\sqrt{2}}\, e^{-i2\pi\nu_L t}$$

The steady-state solution of Eq. (3.2) is given by

$$\mathbf{x}(t) = \mathbf{x}^+(t) + \mathbf{x}^-(t), \qquad \text{where } \mathbf{x}^+(t) = \boldsymbol{\pi}^+\frac{A_0}{\sqrt{2}}\, e^{-i2\pi\nu_L t}\, e^{i\varphi} \tag{3.6}$$

In the above, $\boldsymbol{\pi}^0$, $\boldsymbol{\pi}^+$, and $\boldsymbol{\pi}^-$ represent unit vectors defining the polarization (see Section 3.2 below). Since all physical information is contained in the "half" expressions, the problems can be studied by using these only; the real solutions can then be found by adding the complex conjugate. This definition and procedure is useful because it corresponds to that used in Fourier expansions and transformations [see later, Eq. (3.15) and Eq. (3.16)]. Fourier expansions are used when dealing with the polychromatic case; the integration is taken over negative as well as positive frequencies, and this corresponds to adding the $\mathbf{E}^+$ and $\mathbf{E}^-$ or complex-conjugated terms. Substitution of these expressions into Eq. (3.2) yields the exact solutions:

$$A_0 = \frac{qE_0}{2\pi\nu_L m\gamma} \frac{2\pi\nu_L\gamma}{[16\pi^4(\nu_0^2 - \nu_L^2)^2 + 4\pi^2\nu_L^2\gamma^2]^{1/2}}$$

$$\varphi = \arctan\left[\frac{\nu_L\gamma}{2\pi(\nu_0^2 - \nu_L^2)}\right] \tag{3.7}$$

showing resonance at $\nu_L = \nu_0$.

These exact solutions are clumsy, but they can be simplified for the case of a high-Q resonance, such as is normally the situation in atomic physics. (The Q value of an oscillator is defined as $Q = 2\pi\nu_0/\gamma$.) Then we may place $\nu_L = \nu_0$ in all but the difference $\nu_0 - \nu_L$. The reduction gives the solution

$$\mathbf{x}^+(t) = \boldsymbol{\pi}^+ \frac{qE_0}{2\sqrt{2}\,\pi\nu_0 m\gamma} \frac{\gamma/2}{[4\pi^2(\nu_0 - \nu_L)^2 + \gamma^2/4]^{1/2}} e^{-i2\pi\nu_L t} e^{i\varphi} \tag{3.8}$$

with

$$\varphi = \arctan\left[\frac{\gamma/2}{2\pi(\nu_0 - \nu_L)}\right]$$

These expressions can be written in terms of the Lorentzian given in Eq. (A5.18) of Appendix 5, with $f = (\nu_0 - \nu_L)$ and $T = \gamma/2$:

$$\mathcal{L}(f, T) = \mathcal{L}(\nu_0 - \nu_L, 2/\gamma) = \frac{1}{\{1 + [16\pi^2(\nu_0 - \nu_L)^2/\gamma^2]\}^{1/2}}$$

Then we have

$$\mathbf{x}^+(t) = \boldsymbol{\pi}^+ \frac{qE_0}{2\sqrt{2}\,\pi\nu_0 m\gamma} \mathcal{L}(\nu_0 - \nu_L, 2/\gamma)\, e^{-i2\pi\nu_L t} e^{i\varphi}$$

For the case of a linearly polarized field in the x-direction, $\boldsymbol{\pi}^+ = \boldsymbol{\pi}^- = \mathbf{x}^0$, and the real solution becomes

$$\mathbf{x}(t) = \mathbf{x}^0 \frac{\sqrt{2}\,qE_0}{2\pi\nu_0 m\gamma} \mathcal{L}(\nu_0 - \nu_L, 2/\gamma) \times \cos\left[2\pi\nu_L t - \arctan\frac{\gamma/2}{2\pi(\nu_0 - \nu_L)}\right] \tag{3.9}$$

The rms amplitude is

$$A_0 = \frac{qE_0}{2\pi\nu_0 m\gamma} \frac{\gamma/2}{[4\pi^2(\nu_0 - \nu_L)^2 + \gamma^2/4]^{1/2}}$$

$$= \frac{qE_0}{2\pi\nu_0 m\gamma} \mathcal{L}(\nu_0 - \nu_L, 2/\gamma) \tag{3.10}$$

and the phase is

$$\varphi = \arctan \frac{\gamma}{4\pi(\nu_0 - \nu_L)} \tag{3.11}$$

The use and properties of these functions are discussed in Appendix 5. They are used in some of the analysis that follows.

A similar solution is obtained for a circular polarized force:

$$\mathbf{F}(t) = qE_0[\mathbf{x}^0 \cos(2\pi\nu_L t) + \mathbf{y}^0 \sin(2\pi\nu_L t)]$$

This can be written in the complex notation as

$$\mathbf{F}(t) = \frac{qE_0}{\sqrt{2}} (\boldsymbol{\pi}^+ e^{-i2\pi\nu_L t} + \boldsymbol{\pi}^- e^{+i2\pi\nu_L t}$$

with

$$\boldsymbol{\pi}^+ = \frac{\mathbf{x}^0 + i\mathbf{y}^0}{\sqrt{2}} \quad \text{and} \quad \boldsymbol{\pi}^- = \frac{\mathbf{x}^0 - i\mathbf{y}^0}{\sqrt{2}}$$

The solution of Eq. (3.2) is now

$$\mathbf{x}(t) = A_0[\mathbf{x}^0 \cos(2\pi\nu_L t - \varphi) + \mathbf{y}^0 \sin(2\pi\nu_L t - \varphi)]$$

a circular polarized motion of radial amplitude A_0 and lagging in phase by φ.

In the above equations a factor $\sqrt{2}$ has been included for the case of linear polarization; therefore the "amplitudes" (A_0 and E_0) are the root-mean-square values of the oscillating quantity or the *rms amplitudes.* This usage ensures that any equation for energy, energy density, intensity, or power is independent of the nature of the polarization. The reader must be warned, however, that some equations in this text appear to differ by a factor of 2 from equations quoted in other sources.

It also should be noted that, in general, $\mathbf{E}(t)$ will be the temporal oscillatory factor of a wave:

$$\mathbf{E}(\mathbf{r}, t) = \mathbf{E}^{+}(t)\, e^{+i2\pi\sigma\mathbf{k}^{0}\cdot\mathbf{r}} + \mathbf{E}^{-}(t)\, e^{-i2\pi\sigma\mathbf{k}^{0}\cdot\mathbf{r}}$$

where $\mathbf{k}^0$ is the unit vector defining the propagation direction, and $\sigma = 1/\lambda$ is the wave number. For a linearly polarized plane wave this reduces to

$$\mathbf{E}(\mathbf{r}, t) = \boldsymbol{\pi}^{0}\sqrt{2}\, E_0 \cos 2\pi(\nu_0 t - \sigma\mathbf{k}^{0}\cdot\mathbf{r})$$

3.2. SPECIFICATION OF THE POLARIZATION

In the previous section a notation for specifying the polarization of the radiation has been introduced. Its formal properties are as follows.

An electromagnetic wave, propagating through space in the direction $\mathbf{k}^0$, can be written as

$$\mathbf{E}(\mathbf{r}, t) = \frac{E_0}{\sqrt{2}}[\boldsymbol{\pi}^{+}\, e^{-i2\pi\nu t}\, e^{+i2\pi\sigma\mathbf{k}^{0}\cdot\mathbf{r}} + \boldsymbol{\pi}^{-}\, e^{+i2\pi\nu t}\, e^{-i2\pi\sigma\mathbf{k}^{0}\cdot\mathbf{r}}] \tag{3.12}$$

Associated with the direction of propagation of the wave, we choose a unit vector $\boldsymbol{\pi}^{+}$ orthogonal to $\mathbf{k}^0$, and its complex conjugate $\boldsymbol{\pi}^{-}$, which obey the following relations:

$$\begin{gathered} \mathbf{k}^{0}\cdot\boldsymbol{\pi}^{+} = \mathbf{k}^{0}\cdot\boldsymbol{\pi}^{-} = 0 \\ \boldsymbol{\pi}^{+*} = \boldsymbol{\pi}^{-}, \qquad \boldsymbol{\pi}^{-*} = \boldsymbol{\pi}^{+} \\ \boldsymbol{\pi}^{+}\cdot\boldsymbol{\pi}^{-} = 1 \end{gathered} \tag{3.13}$$

If, for example, we choose $\mathbf{k}^0 = \mathbf{z}^0$, the unit vector along the z axis, we can choose with complete generality

$$\begin{aligned} \boldsymbol{\pi}^{+} &= \mathbf{x}^{0}\cos\alpha + i\mathbf{y}^{0}\sin\alpha \\ \boldsymbol{\pi}^{-} &= \mathbf{x}^{0}\cos\alpha - i\mathbf{y}^{0}\sin\alpha \end{aligned} \tag{3.14}$$

from which is obtained the further relation

$$\boldsymbol{\pi}^{+}\cdot\boldsymbol{\pi}^{+} = \boldsymbol{\pi}^{-}\cdot\boldsymbol{\pi}^{-} = \cos(2\alpha)$$

The angle α is called the *polarization defining parameter.*

Examples. Take $\mathbf{k}^0$ to be along the z axis; the polarization vector will be in the x–y plane:

i. With $\alpha = 0°$, $\boldsymbol{\pi}^+ = \boldsymbol{\pi}^- = \mathbf{x}^0$, and Eq. (3.12) reduces to a linearly polarized field along the x axis:

$$\mathbf{E}(z, t) = \mathbf{x}^0\sqrt{2}\,E_0 \cos 2\pi(\nu t - \sigma z)$$

ii. With $\alpha = 90°$, $\boldsymbol{\pi}^\pm = \pm i\mathbf{y}^0$, and Eq. (3.12) reduces to a linear polarized field along the y axis:

$$\mathbf{E}(z, t) = \mathbf{y}^0\sqrt{2}\,E_0 \sin 2\pi(\nu t - \sigma z)$$

iii. With $\alpha = 45°$, $\boldsymbol{\pi}^\pm = (\mathbf{x}^0 \pm i\mathbf{y}^0)/\sqrt{2}$, and Eq. (3.12) reduces to a circularly (right-hand) polarized field:

$$\mathbf{E}(z, t) = E_0[\mathbf{x}^0 \cos 2\pi(\nu t - \sigma z) + \mathbf{y}^0 \sin 2\pi(\nu t - \sigma z)]$$

With $\alpha = -45°$, left-hand circular polarization is achieved.

iv. The general form of Eq. (3.14) gives elliptical polarization:

$$\mathbf{E}(z, t) = \sqrt{2}\,E_0[\mathbf{x}^0 \cos\alpha \cos 2\pi(\nu t - \sigma z) + \mathbf{y}^0 \sin\alpha \sin 2\pi(\nu t - \sigma z)]$$

Other possible choices of unit vectors $\boldsymbol{\pi}^+$ and $\boldsymbol{\pi}^-$ from which to construct a wave as in Eq. (3.12) are as follows:

a. For a radially propagating spherical wave, $\mathbf{k}^0 = \mathbf{r}^0$. Then one may use unit vectors in the polar and azimuthal directions, $\boldsymbol{\theta}^0$ and $\boldsymbol{\phi}^0$, as illustrated in Fig. 3. Then one can construct

$$\boldsymbol{\pi}^\pm = \boldsymbol{\theta}^0 \cos\alpha \pm i\boldsymbol{\phi}^0 \sin\alpha$$

b. For a wave propagating in the general direction $\mathbf{k}^0$, one may choose any two appropriate orthogonal unit vectors $\mathbf{i}^0$ and $\mathbf{j}^0$.

It can be seen that the general form Eq. (3.12) can be used to describe all directions of propagation and all directions of polarization.

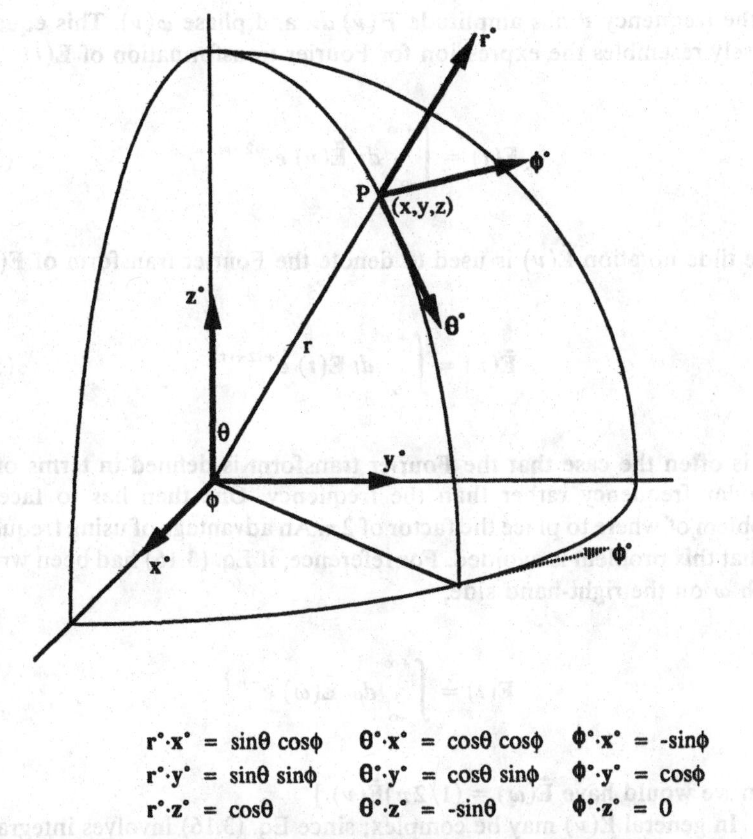

FIGURE 3. The relations between the Cartesian and the spherical coordinate systems.

3.3. POLYCHROMATIC OSCILLATION

The foregoing treatment is restricted to the case of a single frequency (monochromatic radiation). It is appropriate at this point to expand the analysis to the broad-band case. Equation (3.5) may be generalized to

$$\mathbf{E}(t) = \int_0^\infty d\nu \frac{E(\nu)}{\sqrt{2}} (\boldsymbol{\pi}^+ \, e^{-i2\pi\nu t} \, e^{+i\varphi(\nu)} + \boldsymbol{\pi}^- \, e^{+i2\pi\nu t} \, e^{-i\varphi(\nu)}) \quad (3.15)$$

where the frequency component in the infinitesimal frequency window $d\nu$

at the frequency ν has amplitude $E(\nu)\,d\nu$ and phase $\varphi(\nu)$. This equation closely resembles the expression for Fourier transformation of $\mathbf{E}(t)$:

$$\mathbf{E}(t) = \int_{-\infty}^{+\infty} d\nu\, \tilde{\mathbf{E}}(\nu)\, e^{-i2\pi\nu t} \tag{3.16}$$

The tilde notation $\tilde{\mathbf{E}}(\nu)$ is used to denote the Fourier transform of $\mathbf{E}(t)$:

$$\tilde{\mathbf{E}}(\nu) = \int_{-\infty}^{+\infty} dt\, \mathbf{E}(t)\, e^{+i2\pi\nu t} \tag{3.17}$$

[It is often the case that the Fourier transform is defined in terms of the angular frequency rather than the frequency. One then has to face the problem of where to place the factor of 2π. An advantage of using frequency is that this problem is avoided. For reference, if Eq. (3.16) had been written with ω on the right-hand side,

$$\mathbf{E}(t) = \int_{-\infty}^{+\infty} d\omega\, \tilde{\mathbf{E}}(\omega)\, e^{-i\omega t}$$

then we would have $\tilde{\mathbf{E}}(\omega) = (1/2\pi)\tilde{\mathbf{E}}(\nu)$.]

In general $\tilde{\mathbf{E}}(\nu)$ may be complex; since Eq. (3.16) involves integration over positive and negative frequencies, the relationship

$$\tilde{\mathbf{E}}(-\nu) = \tilde{\mathbf{E}}^*(\nu) \tag{3.18}$$

is required to ensure that $\mathbf{E}(t)$ will be real.

The relationship between the differential amplitude $\mathbf{E}(\nu)$ of Eq. (3.15) and the Fourier transform $\tilde{\mathbf{E}}(\nu)$ can be revealed by breaking up the range of integration in Eq. (3.16):

$$\mathbf{E}(t) = \int_{0}^{\infty} d\nu\, \tilde{\mathbf{E}}(\nu)\, e^{-i2\pi\nu t} + \int_{0}^{\infty} d\nu\, \tilde{\mathbf{E}}(-\nu)\, e^{+i2\pi\nu t}$$

$$= \int_{0}^{\infty} d\nu\, [\tilde{\mathbf{E}}(\nu)\, e^{-i2\pi\nu t} + \tilde{\mathbf{E}}^*(\nu)\, e^{+i2\pi\nu t}]$$

Comparison with Eq. (3.15) yields

$$\tilde{\mathbf{E}}(\nu) = \boldsymbol{\pi}^{+} \frac{E(\nu)}{\sqrt{2}} e^{i\varphi(\nu)}$$

$$\tilde{\mathbf{E}}(-\nu) = \tilde{\mathbf{E}}^{*}(\nu) = \boldsymbol{\pi}^{-} \frac{E(\nu)}{\sqrt{2}} e^{-i\varphi(\nu)} \tag{3.19}$$

$$\tilde{\mathbf{E}}(\nu) \cdot \tilde{\mathbf{E}}^{*}(\nu) = \frac{E^2(\nu)}{2}$$

These relationships allow us to use either Eq. (3.15) or (3.16), whichever is more convenient, when discussing broad-band situations. Likewise they allow us to refer (loosely) to the Fourier transform $\tilde{\mathbf{E}}(\nu)$ as the "amplitude" of the component wave of frequency ν. Note, however, that it does not have the physical dimensions of electric field.

3.4. THE EQUATION OF MOTION OF AN OSCILLATING CHARGE

When applying Eq. (3.1) to the case of an oscillating charge the appropriate form of $F_{\text{dissipation}}$ must be found. The rate at which energy is dissipated by radiation of electromagnetic energy is given by Larmor's formula, Eq. (2.9). Multiplying Eq. (3.1) through by the velocity $\dot{\mathbf{x}}$ of the source, an expression for the balance of power is obtained:

$$\dot{K}(t) + P_{\text{rad}} + \dot{U}(t) = P_{\text{applied}} \tag{3.20}$$

where

$$\dot{K}(t) = \frac{d(mv^2/2)}{dt} = m\ddot{\mathbf{x}} \cdot \dot{\mathbf{x}}$$

is the rate of increase of kinetic energy,

$$\dot{U}(t) = \frac{d(kx^2/2)}{dt} = m4\pi^2\nu_0^2\mathbf{x} \cdot \dot{\mathbf{x}}$$

is the rate of increase of potential energy, and

$$P_{\text{applied}} = \mathbf{F}_{\text{applied}} \cdot \dot{\mathbf{x}}$$

is the power of the applied force.

The final term, using Eq. (2.9) with $Q = -e$ appropriate to the case of an electron, gives

$$P_{\text{rad}} = -\mathbf{F}_{\text{dissipation}} \cdot \dot{\mathbf{x}} = -\mathbf{F}_{\text{RR}} \cdot \dot{\mathbf{x}}$$

$$= \frac{\mu_0}{4\pi} \frac{2e^2}{3c} \ddot{\mathbf{x}} \cdot \ddot{\mathbf{x}} = m\tau \ddot{\mathbf{x}} \cdot \ddot{\mathbf{x}} \tag{3.21}$$

where we write

$$\tau = \frac{\mu_0}{4\pi} \frac{2e^2}{3mc} = \frac{2r_0}{3c} \tag{3.22}$$

The quantity τ is a physical constant, the time for light to travel 2/3 of the classical radius, r_0, of the electron. For an electron, $\tau = 6.2664 \times 10^{-24}$ s.

The dissipation force has now been written as $\mathbf{F}_{\text{RR}}$; this is the radiation reaction force, a force reacting back on the oscillator caused by the radiation of energy. Certain problems are immediately apparent. First of all, $\mathbf{F}_{\text{RR}}$ is not itself uniquely defined by Eq. (3.21); for example, it is arbitrary to the extent of any component orthogonal to the velocity $\dot{\mathbf{x}}$. Secondly, difficulty occurs at any instant when $\dot{\mathbf{x}} = 0$ if $\ddot{\mathbf{x}}$ is not also zero; this situation occurs in a linear-polarized oscillator although the difficulty is avoided in a circularly polarized one. Thirdly, the expression used for P_{rad} has included only those terms derived from the radiation field and not those from the inductive field as mentioned at the end of Chapter 2. Presumably the effect of the latter is to distort the simple oscillatory behavior by a perturbation. We can proceed, however, by presuming that the average effect of damping is small and that we can approximate by averaging any small perturbations over a complete cycle. Thus, averaging Eq. (3.21) over a period T yields

$$\langle P_{\text{rad}}(t) \rangle_T = -\frac{1}{T} \int_{t-T}^{t} dt' \, \mathbf{F}_{\text{RR}} \cdot \dot{\mathbf{x}}$$

$$= +\frac{m\tau}{T} \int_{t-T}^{t} dt' \, \ddot{\mathbf{x}} \cdot \ddot{\mathbf{x}}$$

$$= +\frac{m\tau}{T} \dot{\mathbf{x}} \cdot \ddot{\mathbf{x}} \Big|_{t-T}^{t} - \frac{m\tau}{T} \int_{t-T}^{t} dt' \, \dot{\mathbf{x}} \cdot \dddot{\mathbf{x}}$$

If the averaging period is a number of complete cycles of oscillation, the first term on the right-hand side is zero and in any case, as mentioned above, it is zero for the case of a circularly polarized motion when $\dot{\mathbf{x}}$ and $\ddot{\mathbf{x}}$ are orthogonal. Equating the remaining integrands, we obtain the usual expression for the radiation reaction:

$$\mathbf{F}_{\mathrm{RR}} = m\tau\dddot{\mathbf{x}} \tag{3.23}$$

The use of this in Eq. (3.21) is equivalent to replacing Larmor's formula

$$P_{\mathrm{rad}} = m\tau\ddot{\mathbf{x}} \cdot \ddot{\mathbf{x}}$$

by the alternative expression

$$P'_{\mathrm{rad}} = -m\tau\dot{\mathbf{x}} \cdot \dddot{\mathbf{x}} \tag{3.24}$$

It is useful, before proceeding, to explore briefly the differences between P and P'. Consider a charged particle of charge Q which has the equation of motion

$$\mathbf{x}(t) = \boldsymbol{\pi}^{+}\frac{A_0}{\sqrt{2}}e^{-i2\pi\nu t} + \text{c.c.}$$

If we choose $\boldsymbol{\pi}^{+} = \mathbf{x}^0$ this gives linearly polarized motion of amplitude $\sqrt{2}\,A_0$; with the choice $\boldsymbol{\pi}^{+} = \frac{4}{5}\mathbf{x}^0 + i\frac{3}{5}\mathbf{y}^0$ it gives elliptically polarized motion of x amplitude $(4\sqrt{2}/5)A_0$ and y amplitude $(3\sqrt{2}/5)A_0$; and with $\boldsymbol{\pi}^{+} = (1/\sqrt{2})(\mathbf{x}^0 + i\mathbf{y}^0)$ it gives circularly polarized motion of amplitude A_0. Figure 4 illustrates the difference between P_{rad} and P'_{rad} for the three cases. In the linear case the two expressions reduce to

$$P_{\mathrm{rad}} = 32\pi^4\nu^4A_0^2m\tau\cos^2(2\pi\nu t)$$

$$P'_{\mathrm{rad}} = 32\pi^4\nu^4A_0^2m\tau\sin^2(2\pi\nu t)$$

The instantaneous rates of radiation differ by a phase of 180° during a cycle, but the average rates are exactly the same. Providing that the rate of damping is small (which we shall later justify) and that, in consequence, we can regard the details of damping as causing only negligibly small perturbations to the motion, we can proceed by replacing P_{rad} by P'_{rad}, i.e., by using $\mathbf{F}_{\mathrm{dissipation}} = \mathbf{F}_{\mathrm{RR}} = m\tau\dddot{\mathbf{x}}$.

The equation of the motion (3.1) now becomes

$$\ddot{\mathbf{x}} - \tau\dddot{\mathbf{x}} + 4\pi^2\nu_0^2\mathbf{x} = \frac{\mathbf{F}(t)}{m} \tag{3.25}$$

rather than Eq. (3.2) for the frictionally damped case.

The validity of the approximation depends on the size of the dimensionless quantity $\nu_0\tau$ compared with unity. For optical radiation, $\nu_0 \sim 6\times10^{14}$ Hz, and $\nu_0\tau \sim 4\times10^{-9}$. In the characteristic x-ray region, $\nu_0 \sim 3\times10^{18}$ Hz, and $\nu_0\tau \sim 2\times10^{-5}$. Only in the 1-MeV gamma ray region, where $\nu_0 \sim 3\times10^{21}$ Hz and $\nu_0\tau \sim 0.02$ does the approximation begin to fail. We shall proceed by being prepared to neglect terms in $\nu_0\tau$ (or $\nu\tau$) whenever they complicate the analysis.

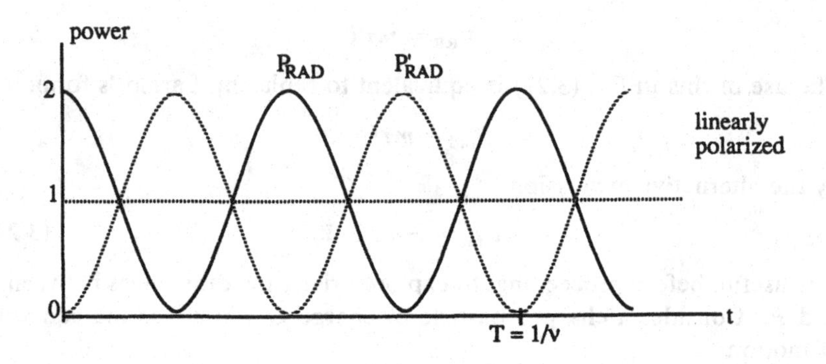

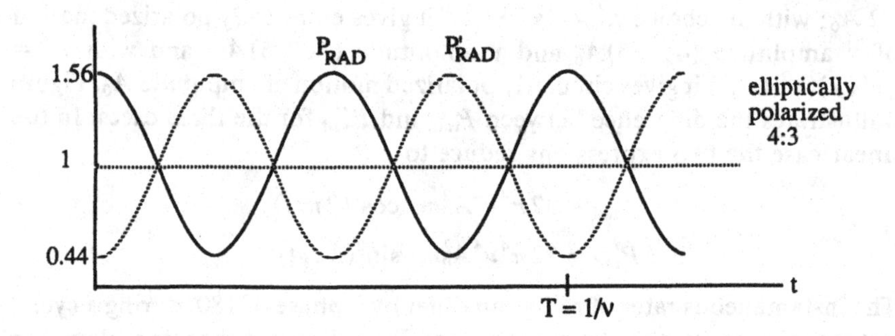

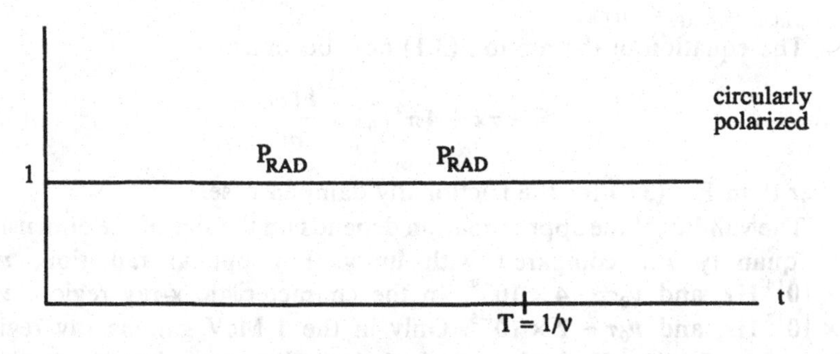

FIGURE 4. The power radiated from an oscillating charge. Power is measured in units of $\langle P\rangle = 16\pi^4\nu_0^4A_0^2m\tau$.

3.5. RADIATION FROM A FREELY OSCILLATING CHARGE

With $\mathbf{F}(t) = 0$, Eq. (3.25) has a solution of the form

$$\mathbf{x}(t) = \boldsymbol{\pi}^{+} \frac{A_0}{\sqrt{2}} e^{-\gamma t/2} e^{-i2\pi\nu' t} + \text{c.c.} \tag{3.26}$$

Substituting this in Eq. (3.25), expressions for the parameters γ and ν' are established. The *classical damping constant* of an oscillating charge is (making a few appropriate approximations)

$$\gamma_{\text{class}} = 4\pi^2 \nu_0^2 \tau = \frac{\mu_0}{4\pi} \frac{8\pi^2 e^2}{3mc} \nu_0^2 \tag{3.27}$$

The frequency of free oscillation is

$$\nu' = \nu_0 (1 - 5\pi^2 \nu_0^2 \tau^2)^{1/2} = \nu_0 \left(1 - \frac{5\gamma^2}{16\pi^2 \nu_0^2}\right)^{1/2} \tag{3.28}$$

This is similar to the expression for the frequency of free oscillation with frictional damping obtained in Eq. (3.4). It differs by the factor 5 in the small second term of the parentheses, which is a consequence of the dissipative term in the equation of motion being proportional to the third derivative of the displacement rather than to the first derivative.

The radiative power deduced from Larmor's formula is

$$P_{\text{rad}}(t) = \frac{\mu_0}{4\pi} \frac{32\pi^4 \nu_0^4 e^2 A_0^2}{3c} e^{-\gamma t} \tag{3.29}$$

in which we have replaced ν' by ν_0. The result is checked by noting that it can be rewritten as

$$P_{\text{rad}}(t) = P_{\text{rad}}(0)\, e^{-\gamma t} = \gamma W_0\, e^{-\gamma t} \tag{3.30}$$

in writing this, the expression for γ given by Eq. (3.27) has been used, and

$$W_0 = 4\pi^2 \nu_0^2 m A_0^2$$

is the energy of an oscillator with rms amplitude A_0. The expression

$$P_{\text{rad}}(0) = \frac{\mu_0}{4\pi} \frac{32\pi^4 \nu_0^4 e^2 A_0^2}{3c} \tag{3.31}$$

for the average radiant power of an oscillator of rms amplitude A_0 is closely related to the Einstein A coefficient, which will be treated later when quantum concepts are introduced in Chapter 8.

3.6. RADIATION FROM A DRIVEN OSCILLATING CHARGE

Let a charge be driven in monochromatic oscillation with displacement

$$\mathbf{x}(t) = \boldsymbol{\pi}^0 \frac{A_0}{\sqrt{2}}[e^{-i2\pi\nu_0 t} + \text{c.c.}] \tag{3.32}$$

Its acceleration is

$$\mathbf{a}(t) = -\boldsymbol{\pi}^0 \frac{4\pi^2\nu_0^2 A_0}{\sqrt{2}}[e^{-i2\pi\nu_0 t} + \text{c.c.}] \tag{3.33}$$

Substituting in Eq. (2.6), this yields for the radiated electric field vector

$$\mathbf{E}^+(\mathbf{r}, t) = +\frac{\mu_0}{4\pi}\frac{QA_0}{r}4\pi^2\nu_0^2\frac{1}{\sqrt{2}}[\mathbf{I} - \mathbf{r}^0\mathbf{r}^0]\cdot\boldsymbol{\pi}^0\, e^{-i2\pi\nu_0(t-r/c)} \tag{3.34}$$

or

$$\mathbf{E}^+(\mathbf{r}, t) = [\mathbf{I} - \mathbf{r}^0\mathbf{r}^0]\cdot\boldsymbol{\pi}^0 kD_0\, e^{-i2\pi\nu_0(t-r/c)} \tag{3.35}$$

where k is an accumulation of physical and other geometric factors and $D_0 = QA_0$ is the classical rms amplitude of the dipole moment.

CHAPTER 4

Scattering of Radiation from a Charge Driven by an Electromagnetic Field

A particle of charge q and mass m, when acted on by an electromagnetic field, is driven into oscillation and, because of the acceleration involved, will therefore radiate. This is seen as radiation scattered by the particle from the incident field. It is the purpose of this chapter to obtain expressions for the cross section of scattering. In order to particularize, the charge will be taken as that of the electron, $q = -e$.

4.1. THE CASE OF THE MONOCHROMATIC FIELD

The electric vector of a monochromatic field of frequency ν_L is represented by

$$\mathbf{E}(t) = \frac{E_0}{\sqrt{2}} (\boldsymbol{\pi}^+ e^{-i2\pi\nu_L t} + \boldsymbol{\pi}^- e^{+i2\pi\nu_L t}) \tag{4.1}$$

Acting on a particle of charge $-e$, this produces an applied force $\mathbf{F}(t) = -e\mathbf{E}(t)$. When this is placed in the equation of motion, Eq. (3.25), appropriate to the case where the damping is supplied by radiation reaction, the steady-state solution is

$$\mathbf{x}(t) = \frac{A(\nu_L)}{\sqrt{2}} (\boldsymbol{\pi}^+ e^{-i2\pi\nu_L t} e^{i\varphi_L} + \boldsymbol{\pi}^- e^{+i2\pi\nu_L t} e^{-i\varphi_L}) \tag{4.2}$$

which oscillates at the same frequency ν_L as the driving field, but lags in phase by $\varphi_L = \varphi(\nu_L)$. On substituting this solution into the equation of motion, expressions for the amplitude and phase are obtained:

$$A(\nu_L) = \frac{-eE_0}{8\pi^3 \nu_L^3 m\tau} \mathscr{R}(\nu_L) \tag{4.3}$$

$$\varphi(\nu_L) = \arctan \frac{2\pi\nu_L^3\tau}{\nu_0^2 - \nu_L^2} \tag{4.4}$$

In Eq. (4.3) the resonant factor $\mathcal{R}$ is given by

$$\mathcal{R}(\nu) = \frac{8\pi^3\nu^3\tau}{[16\pi^4(\nu_0^2 - \nu^2)^2 + 64\pi^6\nu^6\tau^2]^{1/2}} \tag{4.5}$$

The reader is reminded that τ is defined in Eq. (3.22) as

$$\tau = \frac{\mu_0}{4\pi} \frac{2e^2}{3mc} = 6.27 \times 10^{-24} \text{ s}$$

Therefore, at all but extremely high frequencies when classical theory is hardly applicable, the quantity $2\pi\nu\tau$ is a very small quantity; we shall assume that this is the case even when we consider situations when the applied frequency is much larger than the resonance frequency. It may be noted that, on substitution of the expression for $\mathcal{R}$ into that for $A(\nu)$, some factors cancel. However, it is handy to write the expressions in the forms given since the resonant factor $\mathcal{R}$ is then dimensionless and will be used in a number of cases to follow. The variation of $\mathcal{R}(\nu)$ and $\varphi(\nu)$ with frequency are illustrated in Figs. 5a and 5b.

For the normal situation in atomic physics this resonance is very sharp (high Q value), and the expression for $\mathcal{R}$ can be simplified by placing $\nu = \nu_0$ in all but the difference $\nu - \nu_0$. Then, close to resonance, $\mathcal{R}$ becomes the Lorentzian form $\mathcal{L}$ introduced in Section 3.1 and discussed in Appendix 5. With this simplification the results become

$$A(\nu_L) = \frac{-eE_0}{8\pi^3\nu_0^3 m\tau} \mathcal{L}\left(\nu_0 - \nu_L, \frac{1}{2\pi^2\nu_0^2\tau}\right) \tag{4.6}$$

$$\varphi(\nu_L) = \arctan \frac{\pi\nu_0^2\tau}{\nu_0 - \nu_L} \tag{4.7}$$

where

$$\mathcal{L}\left(\nu_0 - \nu, \frac{1}{2\pi^2\nu_0^2\tau}\right) = \frac{2\pi^2\nu_0^2\tau}{[4\pi^2(\nu_0 - \nu)^2 + 4\pi^4\nu_0^4\tau^2]^{1/2}} \tag{4.8}$$

This Lorentzian is appropriate to an oscillator of resonant frequency ν_0 and damping constant γ given by

$$\gamma = \frac{2}{T} = 4\pi^2\nu_0^2\tau$$

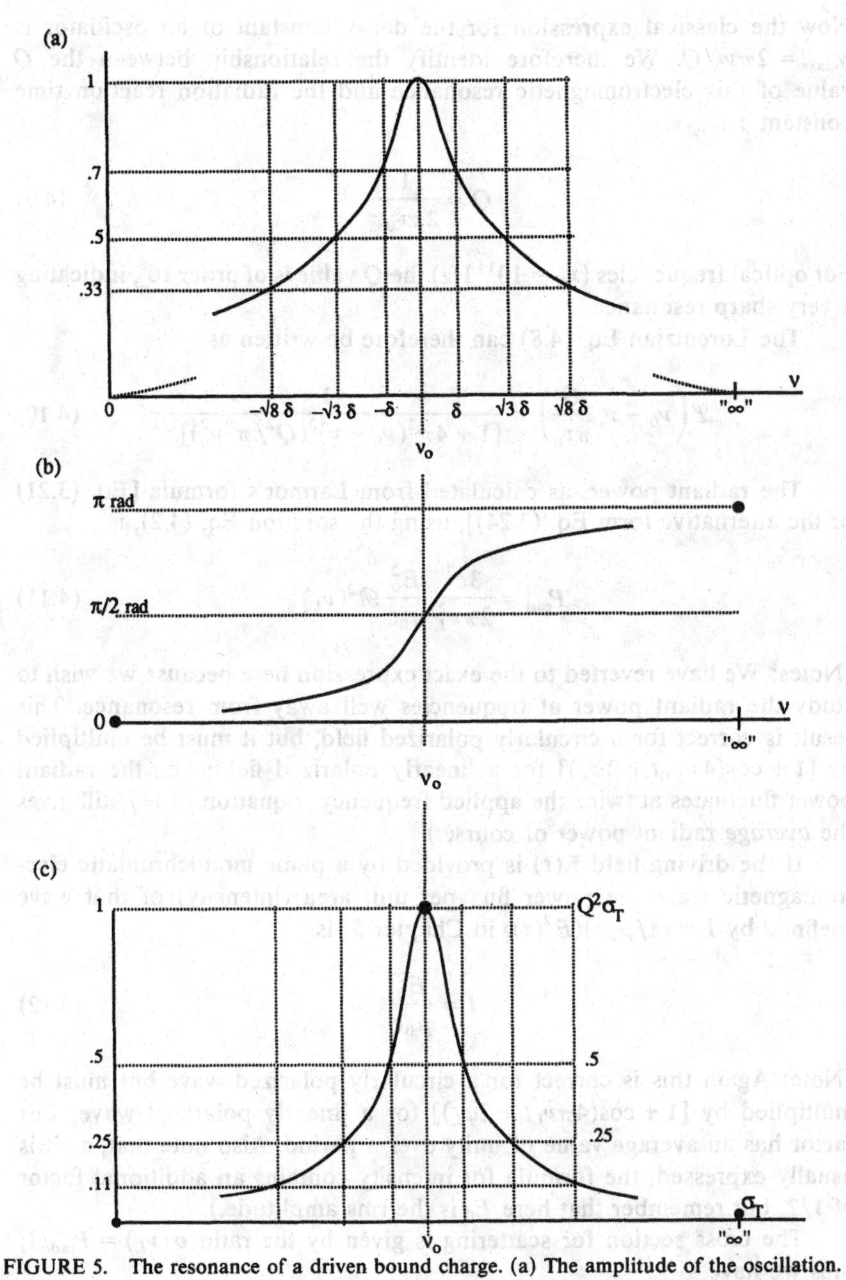

FIGURE 5. **The resonance of a driven bound charge. (a) The amplitude of the oscillation. The frequency shift from resonance $\nu - \nu_0$ is expressed in terms of $\delta = \nu_0/2Q$. (b) The phase of the resonance. (c) The cross section for scattering $\sigma(\nu)$ expressed in terms of the Thompson cross section σ_T.**

Now the classical expression for the decay constant of an oscillator is $\gamma_{\text{class}} = 2\pi\nu_0/Q$. We therefore identify the relationship between the Q value of this electromagnetic resonance and the radiation reaction time constant τ:

$$Q = \frac{1}{2\pi\nu_0\tau} \tag{4.9}$$

For optical frequencies ($\nu_0 \approx 10^{14}$ Hz) the Q value is of order 10^8, indicating a very sharp resonance.

The Lorentzian Eq. (4.8) can therefore be written as

$$\mathscr{L}\left(\nu_0 - \nu, \frac{Q}{\pi\nu_0}\right) = \frac{1}{[1 + 4\pi^2(\nu_0 - \nu)^2(Q^2/\pi^2\nu_0^2)]^{1/2}} \tag{4.10}$$

The radiant power, as calculated from Larmor's formula [Eq. (3.21) or the alternative form Eq. (3.24)], using the solution Eq. (4.2), is

$$P_{\text{rad}} = \frac{3c^2}{2\pi\nu_L^2}\frac{E_0^2}{\mu_0 c}\mathscr{R}^2(\nu_L) \tag{4.11}$$

(Notes: We have reverted to the exact expression here because we wish to study the radiant power at frequencies well away from resonance. This result is correct for a circularly polarized field, but it must be multiplied by $[1 + \cos(4\pi\nu_L t + 2\varphi_L)]$ for a linearly polarized field; i.e., the radiant power fluctuates at twice the applied frequency. Equation (4.11) still gives the *average* radiant power of course.)

If the driving field $\mathbf{E}(t)$ is provided by a plane monochromatic electromagnetic wave the power flux per unit area (intensity) of that wave [defined by $I = (1/\mu_0 c)\langle E^2(t)\rangle$ in Chapter 5] is

$$I = \frac{E_0^2}{\mu_0 c} \tag{4.12}$$

(Note: Again this is correct for a circularly polarized wave but must be multiplied by $[1 + \cos(4\pi\nu_L t + 2\varphi_L)]$ for a linearly polarized wave; this factor has an average value of unity over a period. Also note that, as it is usually expressed, the formula for intensity contains an additional factor of 1/2, but remember that here E_0 is the rms amplitude.)

The cross section for scattering is given by the ratio $\sigma(\nu_L) = P_{\text{rad}}/\text{I}$; thus we have

$$\sigma(\nu_L) = \frac{3c^2}{2\pi\nu_L^2}\mathscr{R}^2(\nu_L) \tag{4.13}$$

The variation of $\sigma(\nu_L)$ with the applied frequency is shown in Fig. 5c.

It is instructive to study this solution in three regimes:

i. At high frequencies: when $\nu_L \gg \nu_0$ (but not so great that it is no longer true that $\nu_L\tau \ll 1$) or alternatively, when $\nu_0 = 0$ (i.e., the case for interaction with the unbound charge). Then $\varphi_L \to \pi$ radian, and

$$\sigma(\nu_L \gg \nu_0) \to 6\pi c^2\tau^2 = \left(\frac{\mu_0}{4\pi}\right)^2 \frac{8\pi e^4}{3m^2} \tag{4.14}$$

This is the *Thomson scattering cross section* for a particle of charge e and mass m. It may be written as

$$\sigma_T = \left(\frac{8\pi}{3}\right) r_0^2$$

where

$$r_0 = \frac{\mu_0}{4\pi}\frac{e^2}{m}$$

is the classical radius. The Thomson cross section is for a free particle but neglects the energy of recoil; in quantum and relativistic language $h\nu_0 \ll h\nu_L \ll mc^2$.

ii. At resonance when $\nu_L = \nu_0$: then $\varphi_L = \pi/2$ radian and

$$\sigma(\nu_L = \nu_0) = \frac{3c^2}{2\pi\nu_L^2} = \sigma_T\left(\frac{1}{2\pi\nu_0\tau}\right)^2 = Q^2\sigma_T \tag{4.15}$$

iii. At low frequencies, $\nu_L \ll \nu_0$: then $\varphi \to 0$ radian and

$$\sigma(\nu_L \ll \nu_0) \to \sigma_T\left(\frac{\nu_L}{\nu_0}\right)^4 = \sigma_R \tag{4.16}$$

This is the region of *Rayleigh scattering* where the cross section is proportional to the fourth power of the frequency. It applies when the frequency of the wave is below the dominant resonant frequency (frequencies) of the scatterer.

4.2. THE CASE OF BROADBAND RADIATION

For the general field, the electric vector may be expressed by the Fourier expansion (Eq. 3.16):

$$\mathbf{E}(t) = \int_{-\infty}^{+\infty} d\nu\, \tilde{\mathbf{E}}(\nu)\, e^{-i2\pi\nu t} \tag{4.17}$$

Note that the monochromatic case of frequency ν_L is recovered by defining

$$\tilde{\mathbf{E}}(\nu) = \frac{E_0}{\sqrt{2}}[\boldsymbol{\pi}^+\delta(\nu - \nu_L) + \boldsymbol{\pi}^-\delta(\nu + \nu_L)] \tag{4.18}$$

where the delta function (see Appendix 5) is defined by the property

$$\int_{-\infty}^{+\infty} dx\, f(x)\delta(x - a) = f(a) \tag{4.19}$$

Since the field is regarded as a superposition of cyclic terms, the use of the radiation reaction, in the form we have derived it, is justified; except that, since Eq. (4.17) extends through frequencies near zero, it may take a very long time for the steady-state cyclic behavior to be established. However, the presence of a natural frequency ν_0 in the equation of motion, Eq. (3.25), ensures that only frequencies in that neighborhood will have significance in the problem.

For the case in which a steady state has been achieved we can expand the displacement by the Fourier integral:

$$\mathbf{x}(t) = \int_{-\infty}^{+\infty} d\nu\, \tilde{\mathbf{x}}(\nu)\, e^{-i2\pi\nu t}$$

Finding the time differentials of $\mathbf{x}(t)$ and placing them in the equation of motion, we obtain

$$\int_{-\infty}^{+\infty} d\nu(-4\pi^2\nu^2 - i8\pi^3\nu^3\tau + 4\pi^2\nu_0^2)\tilde{\mathbf{x}}(\nu)\, e^{-i2\pi\nu t} = \frac{-e}{m}\int_{-\infty}^{+\infty} d\nu\, \tilde{\mathbf{E}}(\nu)\, e^{-i2\pi\nu t}$$

A necessary condition for a steady-state solution is that the coefficients of each frequency component must be the same on each side of the equation. Therefore, equating the integrands, we obtain

$$\tilde{\mathbf{x}}(\nu) = \frac{-e\tilde{\mathbf{E}}(\nu)}{8\pi^3\nu^3 m\tau}\mathcal{R}(\nu)\, e^{i\varphi(\nu)}$$

$$\varphi(\nu) = \arctan\frac{2\pi\nu^3\tau}{\nu_0^2 - \nu^2} \tag{4.20}$$

The result is in formal agreement with the results of the monochromatic case quoted in Eqs. (4.3) and (4.4). The radiant power is obtained from Eq. (3.21):

$$P_{\text{rad}}(t) = -\mathbf{F}_{\text{RR}} \cdot \dot{\mathbf{x}} = m\tau\ddot{\mathbf{x}}(t) \cdot \ddot{\mathbf{x}}(t) \tag{4.21}$$

In this case, because of the spectral width, it is necessary to average over many periods to achieve an average value. The methods by which this is carried out are discussed in Appendix 1. We shall be concerned only with the steady-state result which is achieved by taking the averaging period, or the integration time constant T to approach infinity. The result is

$$P_{\text{rad}} = \langle P_{\text{rad}}(t)\rangle_{T\to\infty}$$

$$= \int_0^{+\infty} d\nu \lim_{T\to\infty} \frac{\tilde{\mathbf{E}}(\nu)\cdot\tilde{\mathbf{E}}^*(\nu)}{\mu_0 c T} \frac{3c^2}{2\pi\nu^2}\mathcal{R}^2(\nu) \tag{4.22}$$

The first factor in the integrand is the *spectral intensity*, the power per unit area per unit frequency interval. It is shown in Appendix 1, Eqs. (A1.18) and (A1.19) that

$$s(\nu) = \lim_{T\to\infty} \frac{\tilde{\mathbf{E}}(\nu)\cdot\tilde{\mathbf{E}}^*(\nu)}{\mu_0 c T} \tag{4.23}$$

and that this is related to the *intensity* of the radiation field by

$$I = \int_0^{\infty} d\nu\, s(\nu) = \int_0^{\infty} d\nu \lim_{T\to\infty} \frac{\tilde{\mathbf{E}}(\nu)\cdot\tilde{\mathbf{E}}^*(\nu)}{\mu_0 c T} \tag{4.24}$$

a result that is to be compared with Eq. (4.12) for a monochromatic wave. The quantities *intensity* and *spectral intensity* are further discussed and defined in Chapter 5.

The remaining factor in the integrand of Eq. (4.22) is identical to the cross section $\sigma(\nu)$ already deduced for monochromatic radiation, Eq. (4.13). Thus we can rewrite Eq. (4.22) as

$$P_{\text{rad}} = \int_0^{\infty} d\nu\, s(\nu)\sigma(\nu) \tag{4.25}$$

CHAPTER 5

Intensity, Energy Density, the Poynting Vector, and Their Spectral Distributions

5.1. THE INTENSITY OF THE RADIATION FIELD

In the previous chapter a quantity called "intensity of radiation" was introduced when deducing expressions for the scattering cross section. In Eq. (4.12), for example, we introduced the intensity $I = E_0^2/\mu_0 c$ appropriate to a plane monochromatic wave; it is the power flux through unit area placed perpendicular to the propagation direction of the wave. (This quantity is more properly called the *irradiance*, but we shall use here the more common description.)

Then, through Eqs. (4.23) and (4.24), we introduced a *spectral intensity* $s(\nu)$, and a *total intensity I*. If this were to represent the situation in a broad-band *beam* (i.e., a unidirectional flow) of radiation, it could also be described as the power flux through unit area. However, such a description for intensity is too restrictive. We notice, to begin with, that neither Eq. (4.23) nor Eq. (4.24) contains any reference to *direction* of energy flow. In practice, at least where the transfer of momentum from the radiation to the scattering particle is not an issue, it is the square of the incident electric field, $\mathbf{E}(t) \cdot \mathbf{E}(t)$ (and perhaps its history up to the moment of time at which we are concerned), that determines the rate of radiation from a scattering charge. The scattered radiation itself is insensitive to the *direction* of energy flow in the incoming field; indeed, that field may be stationary, as in the cavity of a Fabry–Perot interferometer, or in a constant temperature enclosure. We therefore adopt the definition of intensity to be "the time-averaged *energy per unit volume* in the radiation field at the point in question, *multiplied by the velocity of light*":

$$I(t) = c\langle u(t)\rangle_T \tag{5.1}$$

where $u(t)$ is the energy per unit volume in the radiation field. The Poynting vector, $\mathbf{S}(t)$ of Eq. (2.8), can be used for cases where the flow direction is important; $I(t)$ and $\mathbf{S}(t)$ are, of course, dimensionally equivalent.

5.2. ENERGY DENSITY AND THE POYNTING VECTOR

The energy density of an electric field is $u_E(t) = (\varepsilon_0/2)\mathbf{E}(t) \cdot \mathbf{E}(t)$. This is a particular case of a more general quantity $u_E(\mathbf{x}_1, t_1; \mathbf{x}_2, t_2)$ used in describing the coherence properties of the radiation field; it is called the *mutual coherence function* or the *cross-correlation function* of the fields $\mathbf{E}(\mathbf{x}_1, t_1)$ and $\mathbf{E}(\mathbf{x}_2, t_2)$ at two field points of coordinate positions $\mathbf{x}_1$ and $\mathbf{x}_2$, and at two times t_1 and t_2, respectively. When the two points coincide $(\mathbf{x}_1 = \mathbf{x}_2)$, we then speak of the *self-coherence* or the *autocorrelation* of the field $u_E(t_1, t_2) = (\varepsilon_0/2)\mathbf{E}(t_1) \cdot \mathbf{E}(t_2)$. This reduces to the energy density at a field point at a time t when $t_1 = t_2$.

The oscillating electric field which produces the interaction with a charged particle or with an atom is supplied by electromagnetic radiation passing through the field point where the particle or atom resides. According to electromagnetic theory there is an associated oscillating magnetic field $\mathbf{B}(\mathbf{x}, t) = [\mathbf{k}^0 \times \mathbf{E}(\mathbf{x}, t)]/c$ at that point, where $\mathbf{k}^0$ is the unit vector of wave propagation. The energy density of the magnetic field is $u_B(t) = (1/2\mu_0)\mathbf{B}(t) \cdot \mathbf{B}(t)$. Electromagnetic theory shows that (in vacuum) u_E and u_B are equal in magnitude. The *total energy density* in the field is therefore given by

$$u(t) = u_E(t) + u_B(t) = 2u_E(t) = \varepsilon_0 \mathbf{E}(t) \cdot \mathbf{E}(t) \tag{5.2}$$

It is instructive to consider the variations in time and space of these quantities $(\mathbf{E}, \mathbf{B}, \mathbf{S}, u, I)$ for a few simple cases involving monochromatic plane waves.

i. A Progressive Wave, Linearly Polarized. (We revert, for simplicity of writing, to the use of $\omega = 2\pi\nu$, and $k = 2\pi\nu/c = 2\pi/\lambda$, and introduce $\psi_+ = (\omega t - kz)$ for the phase of a wave propagating along the z axis in the positive sense)

$$\mathbf{E}(\mathbf{z}, t) = \mathbf{x}^0\sqrt{2}\,E_0 \cos(\omega t - kz) = \mathbf{x}^0\sqrt{2}\,E_0 \cos\psi_+$$

$$\mathbf{B}(\mathbf{z}, t) = \mathbf{y}^0\sqrt{2}\,\frac{E_0}{c}\cos\psi_+$$

$$u_E(z, t) = u_B(z, t) = \varepsilon_0 E_0^2 \cos^2\psi_+ = \frac{E_0^2}{\mu_0 c^2}\cos^2\psi_+ \tag{5.3}$$

$$u(z, t) = \frac{E_0^2}{\mu_0 c^2}(1 + \cos 2\psi_+)$$

$$\mathbf{S}(z, t) = \mathbf{z}^0\frac{E_0^2}{\mu_0 c}(1 + \cos 2\psi_+)$$

At all points along the wave $u_E = u_B$, but both vary in time and space as $\cos^2 \psi_+$. This factor, when averaged over a whole number of periods (or wavelengths), has the value 1/2. Therefore we have

$$\langle u_E \rangle = \langle u_B \rangle = \frac{E_0^2}{2\mu_0 c^2}$$

$$\langle u \rangle = \langle u_E + u_B \rangle = \frac{E_0^2}{\mu_0 c^2}$$

$$I = c\langle u \rangle = \frac{E_0^2}{\mu_0 c}$$

(Remember that E_0 is the rms amplitude; written in terms of peak amplitude $\langle u \rangle = E_{\mathrm{pk}}^2/2\mu_0 c$, for example.) And, for the Poynting vector we have

$$\langle \mathbf{S} \rangle = \mathbf{z}^0 \frac{E_0^2}{\mu_0 c} = \mathbf{z}^0 c\langle u \rangle$$

It is seen that the intensity I is equal to the magnitude of the (time-averaged) Poynting vector. We repeat, however, that the intensity is more appropriately defined as c times that average energy per unit volume; it can then be used for stationary fields.

ii. A Progressive Wave, Circularly Polarized.

$$\mathbf{E}(\mathbf{z}, t) = E_0(\mathbf{x}^0 \cos \psi_+ + \mathbf{y}^0 \sin \psi_+)$$

$$\mathbf{B}(\mathbf{z}, t) = \frac{E_0}{c}(-\mathbf{x}^0 \sin \psi_+ + \mathbf{y}^0 \cos \psi_+)$$

$$u_E = u_B = \frac{E_0^2}{2\mu_0 c^2}$$

$$u = \frac{E_0^2}{\mu_0 c^2} \tag{5.4}$$

$$I = cu = \frac{E_0^2}{\mu_0 c}$$

$$\mathbf{S} = \mathbf{z}^0 \frac{E_0^2}{\mu_0 c} = \mathbf{z}^0 cu$$

It is to be noted that, in a circularly polarized wave, there is no temporal or spatial variation in u_E, u_B, u, and $\mathbf{S}$, so that averaging is not necessary. The consequence of this fact is that it is often easier, when discussing the interaction of an electromagnetic wave with a charged particle, to use a circularly polarized wave as the simplest mode or state of the radiation field. Linear polarization can then be thought of as a superposition of two opposite circular modes. The circular polarized description also reminds us of the fact that emission or absorption of light is always accompanied by a change of angular momentum.

iii. A Standing Wave, Linearly Polarized. This is produced by a wave such as in Eq. (5.3) superimposed on a similar counterpropagating wave, i.e., with a phase $\psi_- = (\omega t + kz)$, such as would be produced by a reflection of the positively traveling wave from the x–y plane at $z = 0$. With the boundary condition $\mathbf{E}(z = 0, t) = 0$ we have

$$\begin{aligned}
\mathbf{E}(z, t) &= \mathbf{x}^0\sqrt{2}\,E_0 \cos\psi_+ - \mathbf{x}^0\sqrt{2}\,E_0\cos\psi_- \\
&= \mathbf{x}^0 2\sqrt{2}\,E_0 \sin(kz)\sin(\omega t) \\
\mathbf{B}(z, t) &= \mathbf{y}^0\sqrt{2}\frac{E_0}{c}\cos\psi_+ + \mathbf{y}^0\sqrt{2}\frac{E_0}{c}\cos\psi_- \\
&= \mathbf{y}^0 2\sqrt{2}\frac{E_0}{c}\cos(kz)\cos(\omega t) \\
u_E(z, t) &= 4\frac{E_0^2}{\mu_0 c^2}\sin^2(kz)\sin^2(\omega t) \\
u_B(z, t) &= 4\frac{E_0^2}{\mu_0 c^2}\cos^2(kz)\cos^2(\omega t) \\
\mathbf{S}(z, t) &= \mathbf{z}^0 2\frac{E_0^2}{\mu_0 c}\sin(2kz)\sin(2\omega t)
\end{aligned} \tag{5.5}$$

It will be noted that u_E and u_B are no longer equal at every point and at every instant, although their average values are. Averaged over whole wavelengths (or periods),

$$\langle u\rangle = \langle u_E(z, t) + u_B(z, t)\rangle = 2\frac{E_0^2}{\mu_0 c^2}$$

$$I = c\langle u\rangle = 2\frac{E_0^2}{\mu_0 c}$$

$$\langle\mathbf{S}\rangle = 0$$

As is to be expected for a standing wave, derived from two waves transporting energy in opposite senses, the average value of the Poynting vector is zero, although the intensity of the field is not. On the spatial scale of one wavelength and the temporal scale of one period, energy *is* flowing; this is illustrated in Fig. 6a, where we see that energy is interchanging between the electric and magnetic fields which have their peak values at different points in space and at different times. Because $\langle \mathbf{S} \rangle = 0$ it is not a useful quantity for specifying the strength of the interaction between the radiation and charged particles. In this case, it is more appropriate to use the *intensity* I as introduced above. This is effectively a superposition of an average flow of energy per unit area per unit time of $E_0^2/\mu_0 c$ in one sense, and $E_0^2/\mu_0 c$ in the opposite sense (see the results in case i above).

iv. A Standing Wave, Circularly Polarized.

$$\mathbf{E}(z, t) = E_0(\mathbf{x}^0 \cos \psi_+ + \mathbf{y}^0 \sin \psi_+) - E_0(\mathbf{x}^0 \cos \psi_- + \mathbf{y}^0 \sin \psi_-)$$

$$= 2E_0 \sin(kz)[\mathbf{x}^0 \sin(\omega t) - \mathbf{y}^0 \cos(\omega t)]$$

$$\mathbf{B}(z, t) = 2\frac{E_0}{c} \cos(kz)[-\mathbf{x}^0 \sin(\omega t) + \mathbf{y}^0 \cos(\omega t)]$$

$$u_E(z) = 2\frac{E_0^2}{\mu_0 c^2} \sin^2(kz)$$

$$u_B(z) = 2\frac{E_0^2}{\mu_0 c^2} \cos^2(kz) \tag{5.6}$$

$$u = 2\frac{E_0^2}{\mu_0 c^2}$$

$$I = cu = 2\frac{E_0^2}{\mu_0 c}$$

$$\mathbf{S} = 0$$

It will be noted that, although u_E and u_B vary in space, they are constant in time and u is constant in both time and space. (The situation is illustrated in Fig. 6b.) The Poynting vector is always zero. Again, therefore, the intensity is appropriately defined from the energy density.

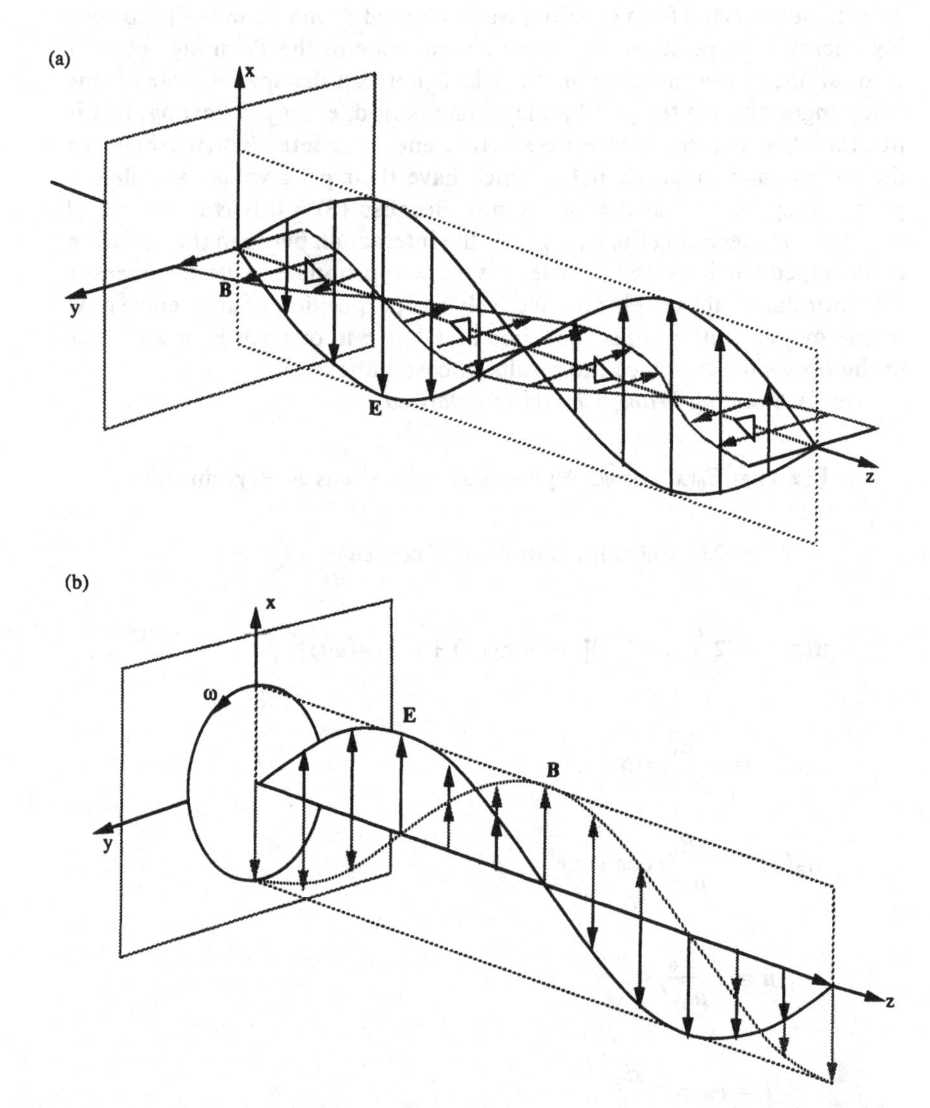

FIGURE 6. Stationary electromagnetic waves; produced by a wave along the z axis superimposed by its reflection from an x–y plane. (a) Linearly polarized: the **E** and **B** fields oscillate back and forth between the limits shown. The **B** field, as shown by the shaded arrows, is drawn at $t = 0$, at which time $\mathbf{E} = 0$ everywhere. The **E** field, as shown by the solid arrows, is drawn at $t = T/4$, at which time $\mathbf{B} = 0$ everywhere. The instantaneous energy flow (Poynting vector) is shown by the open arrows, drawn at $t = T/8$; the flows reverse in direction every $T/4$. (b) Circularly polarized: the **E** and **B** fields remain constant in magnitude at any position z, but rotate with angular frequency ω. There is no energy flow in the field.

All examples confirm the appropriateness of defining intensity through Eq. (5.1) rather than through the Poynting vector.

5.3. SPECTRAL DISTRIBUTIONS

$I(t)$ and $\langle u(t)\rangle$ of Eq. (5.1) can both be expanded in terms of spectral distributions:

$$I(t) = \int_0^\infty d\nu\, s(\nu, t) \tag{5.7}$$

$$\langle u(t)\rangle_T = \int_0^\infty d\nu \langle \rho(\nu, t)\rangle_T \tag{5.8}$$

with

$$s(\nu, t) = c\langle \rho(\nu, t)\rangle_T \tag{5.9}$$

which relates the ***spectral intensity*** $s(\nu, t)$ to the ***spectral energy density*** $\langle \rho(\nu, t)\rangle_T$. The residual time dependence of $s(\nu, t)$ and $I(t)$ vanish as the time-averaging interval T is extended to infinity; they are then written as steady-state quantities $s(\nu)$ and I as in Eqs. (4.23) and (4.24), respectively.

It remains for us to establish an expression for $s(\nu)$ and to justify Eq. (4.23). The energy density in the field is given by Eq. (5.2), $u(t) = \varepsilon_0 \mathbf{E}(t) \cdot \mathbf{E}(t)$. The method of expanding this over frequency, and of performing a time average over an interval T, is discussed in Appendix 1. The result is given in general by Eq. (A1.11):

$$\langle u(t)\rangle_T = \int_{-\infty}^{+\infty} d\nu \int_{-\infty}^{+\infty} df\, \varepsilon_0 \tilde{\mathbf{E}}(\nu) \cdot \tilde{\mathbf{E}}^*(\nu + f)\, e^{i2\pi f t} \mathscr{L}(f, T)\, e^{-i\lambda(f)} \tag{5.10}$$

where the amplitude function $\mathscr{L}(f, T)$ and the phase function $\lambda(f)$ have forms that depend on how the averaging over time is performed; here they are written as Lorentzian, implying that exponential averaging has been used. The frequency $f = \nu' - \nu$, the difference between two frequencies in the radiation field. As shown in Fig. 32 (Appendix 5), $\mathscr{L}(f, T)$ has a significant value only for $fT \lesssim 1$; therefore only frequencies $f \lesssim 1/T$ are of any significance in Eq. (5.10). As discussed in the examples in Appendix 1, such frequencies constitute the residual noise after averaging a signal over an integration time constant T, or they can be the beat-frequencies between the coherent superposition of frequencies in the radiation field.

The latter occurs, for example, in "time-resolved spectroscopy" or "light-beats" experiments. The time constant T in such experiments must be chosen short enough that the beat frequency is retained in the signal, but long enough to average effectively over rapid fluctuations. In the mathematical description of such experiments by the use of Eq. (5.10) the averaging time T must be chosen short enough that the desired frequencies are retained.

When the time constant tends to infinity, the *steady-state* signal is approached. As shown in Appendix 1, this yields for the spectral energy density distribution of Eq. (5.9)

$$\rho(\nu) = \lim_{T\to\infty} \frac{\tilde{\mathbf{E}}(\nu)\cdot\tilde{\mathbf{E}}^*(\nu)}{\mu_0 c^2 T} \tag{5.11}$$

so that

$$s(\nu) = c\rho(\nu) = \lim_{T\to\infty} \frac{\tilde{\mathbf{E}}(\nu)\cdot\tilde{\mathbf{E}}^*(\nu)}{\mu_0 c T} \qquad \text{(QED)} \tag{5.12}$$

This analysis could have been carried out using the spectral amplitudes $\mathbf{E}(\nu)$ of Eq. (3.16) rather than using the Fourier transforms $\tilde{\mathbf{E}}(\nu)$ of Eq. (3.17). The final result, using Eq. (3.19), would then be

$$s(\nu) = c\rho(\nu) = \lim_{T\to\infty} \frac{E^2(\nu)}{2\mu_0 c T} \tag{5.13}$$

Equation (5.12), or its alternative form Eq. (5.13), for the spectral intensity is expressed in terms of quantities that are not directly measurable. The quantity $s(\nu)\delta\nu$ is, in principle, a measurable quantity, being the power per unit area measured in a spectral window of width $\delta\nu$ at the frequency ν and averaged over an observing time T. It is therefore appropriate to rewrite Eq. (5.13) in the form

$$s(\nu)\delta\nu = \frac{1}{T\delta\nu}\frac{|E(\nu)\delta\nu|^2}{2\mu_0 c} \tag{5.14}$$

where the limit to infinity has been removed but it is understood that this limit may be required in order to make the expression accurate. The quantity $E(\nu)\delta\nu$ is the rms amplitude of the field in the frequency window $\delta\nu$. For optical fields this may not, in practice, be observable, either because the frequency is too high for present technology or because there are some as yet undiscussed effects (quantal) which preclude its measurement. But certainly it can be measured at lower frequencies. We shall adopt the attitude that, in principle, $E(\nu)\delta\nu$ is measurable. The two other factors in Eq. (5.14), T and $\delta\nu$, would seem to be independent of each other, but, from an experimental point of view, they are related by an optimal condition. When averaging over a period T one removes from the signal all beat, modulation, and noise frequencies higher than $\Delta\nu = 1/T$. If $\Delta\nu < \delta\nu$, one is removing, by the time-averaging process, frequencies that have been allowed to pass through the chosen size of the frequency window of the experiment, $\delta\nu$. If $\Delta\nu > \delta\nu$, then the window has already removed frequencies which could have been observed by the time-averaging process. Obviously the condition $\Delta\nu \approx \delta\nu$ or $T\delta\nu \approx 1$ optimizes the efficiency of the system. The condition gives an operational meaning to the limiting process expressed in Eqs. (5.12) and (5.13).

CHAPTER 6

THE INTERACTION OF A BEAM OF ELECTROMAGNETIC RADIATION WITH A FREE ELECTRIC CHARGE

THE COMPTON EFFECT

We have studied the classical description of an electromagnetic radiation field and of its interaction with a charge. It is now necessary to study a simple case, both theoretically and experimentally, to find out whether there are further restrictions and rules that govern the interaction. We choose for study the simplest case, the *scattering of radiation by a free charge*; the natural frequency ν_0 of Eq. (3.2) is zero. The damping will also be neglected. However, we now recognize that *linear momentum* will be transferred to the charge through the interaction of the magnetic field of the radiation with the movement of the charge. Firstly we shall look at the experimental facts and then we shall compare this with the theoretical prediction. The physical phenomenon goes under the name of the Compton effect.

6.1. THE COMPTON EFFECT

Experimentally it has been observed that when an electromagnetic wave of frequency ν_0 is scattered by a free electron at an angle θ, the frequency of the scattered radiation is given by

$$\nu = \frac{\nu_0}{1 + \varepsilon(1 - \cos\theta)} \tag{6.1}$$

The experiments studying this phenomenon were initially performed by A. H. Compton, and such scattering is called Compton scattering. In his experiments the electrons were not strictly free, being bound to atoms in the scattering center. The frequency chosen for the incident radiation was sufficiently high ($\nu_0 = 4.2 \times 10^{18}$ Hz, $\lambda = 0.07$ nm being the K_α characteristic x-radiation of molybdenum) that the electrons in the atoms of the scatterer (a graphite block) could be regarded as effectively free. Even so an "unmodified" scattered line of frequency ν_0 was observed, showing that some of the scattering was from *bound* electrons. We concentrate our attention on the Compton scattering, however.

Experiments using different incident frequencies establish that the empirical constant ε in Eq. (6.1) is proportional to ν_0, the frequency of the radiation; therefore we can write $\varepsilon = \tau_C \nu_0$. Since ε is dimensionless, τ_C is a characteristic time associated with the process. Its value is determined experimentally to be $\tau_C = 8.0933 \times 10^{-21}$ s, and may be called the *Compton period of the electron.* Compton expressed this as a wavelength, the Compton wavelength of the electron, $\lambda_C = c\tau_C = 2.4 \times 10^{-12}$ m. And, of course, the accuracy he achieved was not that implied by the figure quoted for τ_C above; the value quoted anticipates later theory and experiment. With the use of these quantities, Compton's scattering equation would be written in its more usual form:

$$\lambda = \lambda_0 + \lambda_C(1 - \cos\theta)$$

It can also be anticipated that, should experiments be performed using other charged particles than electrons, the empirical value measured for ε (and for τ_C) would be inversely proportional to the mass of the particle used (in practice the experiments have not been done). We can therefore write

$$\varepsilon = \tau_C \nu_0 = \frac{\kappa \nu_0}{m_0} \quad \text{or} \quad \kappa = \tau_C m_0 \tag{6.2}$$

where m_0 is the rest mass of the charged particle. The value of the constant of proportionality can now be determined: $\kappa = 7.3726 \times 10^{-51}$ kg s. Since we have been factoring out particular physical constants appropriate to the scattering particle being used, we perceive the emergence of a physical constant κ. But we must now turn to theory for guidance.

6.2. A CLASSICAL THEORY OF THE COMPTON EFFECT

For the incident radiation choose a simple monochromatic electromagnetic plane wave, circularly polarized, propagating in vacuum along the z axis (recall the earlier statement, made after Eq. (5.4), that it is often simpler to deal with circularly polarized light). In the laboratory (L) frame, this is radiation is described by the equations

$$\mathbf{E}(z, t) = E_0(\mathbf{x}^0 \cos \psi + \mathbf{y}^0 \sin \psi)$$

$$\mathbf{B}(z, t) = \frac{E_0}{c}(-\mathbf{x}^0 \sin \psi + \mathbf{y}^0 \cos \psi) \tag{6.3}$$

where the phase $\psi = 2\pi\nu_0(t - z/c)$. The Poynting vector for the incident wave is

$$\mathbf{S}_0 = \frac{\mathbf{E} \times \mathbf{B}}{\mu_0} = \mathbf{z}^0 \frac{E_0^2}{\mu_0 c} \tag{6.4}$$

The wave is to interact with a particle of charge q and rest mass m initially at rest in the laboratory. The initial motion of the charged particle will obviously be complicated under the combined action of the electric and magnetic fields. We shall seek a *steady-state solution.* The rotating $\mathbf{E}$ field guides the particle toward a circular path, and the interaction between this circular velocity and the $\mathbf{B}$ field of the radiation pushes the particle "down stream," i.e., forward in the direction of propagation of the incident radiation. This is a case of transfer of momentum to the charged particle which hitherto we have neglected, and is the primary reason why we are studying this case. Initially one may be inclined to think that this should lead to some sort of continuous acceleration—a continuous transfer of energy from the field to the charge. This point, which is at variance with the observed facts, has led to the belief that one cannot properly describe the Compton effect by classical interaction, and has supported the view that the Compton effect demands quantization of the electromagnetic field. A little reflection, however, reminds us that, in many cases in classical physics, the system responds to a periodic applied stimulus and reaches a new steady state.

We therefore seek a steady-state solution in which there is a *constant down-stream velocity* $\mathbf{V} = \mathbf{z}^0\beta c$, on which is superimposed a constant rotation at the frequency of the passing wave. The superposition of these two motions thereby produces a helical path for the classical description of the motion of the charged particle after it has interacted with the radiation.

Let us suppose that such a steady-state motion exists. Transform to a new frame of reference moving with velocity $\mathbf{V}$ with respect to the laboratory. This is the zero-momentum (ZM) frame, because after interaction, when the steady state is reached, the particle has zero linear momentum in it. Quantities measured in the ZM frame are labeled by an asterisk. The Lorentz transformations, between the laboratory and the zero-momentum frames, for the vectors and frequency of the wave are

$$E^* = E_0\left(\frac{1-\beta}{1+\beta}\right)^{1/2}$$

$$B^* = B_0\left(\frac{1-\beta}{1+\beta}\right)^{1/2} \tag{6.5}$$

$$\nu^* = \nu_0\left(\frac{1-\beta}{1+\beta}\right)^{1/2}$$

The transformations of the fields are appropriate to the case where the fields are transverse to the velocity of transformation; the transformation of the frequency is the Doppler effect for a wave traveling along the direction of the velocity of transformation. The phase of the wave is invariant:

$$\psi^* = 2\pi\nu^*(t^* - z^*/c) = 2\pi\nu(t - z/c) = \psi \tag{6.6}$$

Therefore, in the ZM frame, the equations of the wave are

$$\mathbf{E}^*(z^*, t^*) = E^*(\mathbf{x}^0 \cos\psi + \mathbf{y}^0 \sin\psi)$$

$$\mathbf{B}^*(z^*, t^*) = \frac{E^*}{c}(-\mathbf{x}^0 \sin\psi + \mathbf{y}^0 \cos\psi) \tag{6.7}$$

$$\mathbf{S}^* = \mathbf{z}^0\frac{E^{*2}}{\mu_0 c} = \mathbf{S}_0\frac{1-\beta}{1+\beta}$$

It is straightforward to recognize a steady state in the ZM frame; it is illustrated in Fig. 7a. The (positively) charged particle is rotating in antiphase to $\mathbf{E}^*$ under the centripetal force $\mathbf{F}^* = q\mathbf{E}^*$. Its velocity $\mathbf{v}^*$, tangential to the circular path, is always antiparallel to $\mathbf{B}^*$ so there is no magnetic force to further accelerate it downstream. The very small radiation-reaction force has not been included. Also neglected is the force $\mathbf{F} = \nabla\mathbf{B} \cdot \mathbf{m}$ produced by the gradient of the magnetic field acting on the magnetic moment $\mathbf{m}$ of the charged particle. The particle rotates in a circle at a fixed value of z^* (i.e., with a constant velocity $\mathbf{z}^0 V$ in the L frame), and with angular velocity $\omega^* = 2\pi\nu^*$ in synchronism with the passing circularly polarized wave of frequency ν^*. The centripetal equation

$$4\pi^2\nu^{*2}m^*R^* = qE^* \tag{6.8}$$

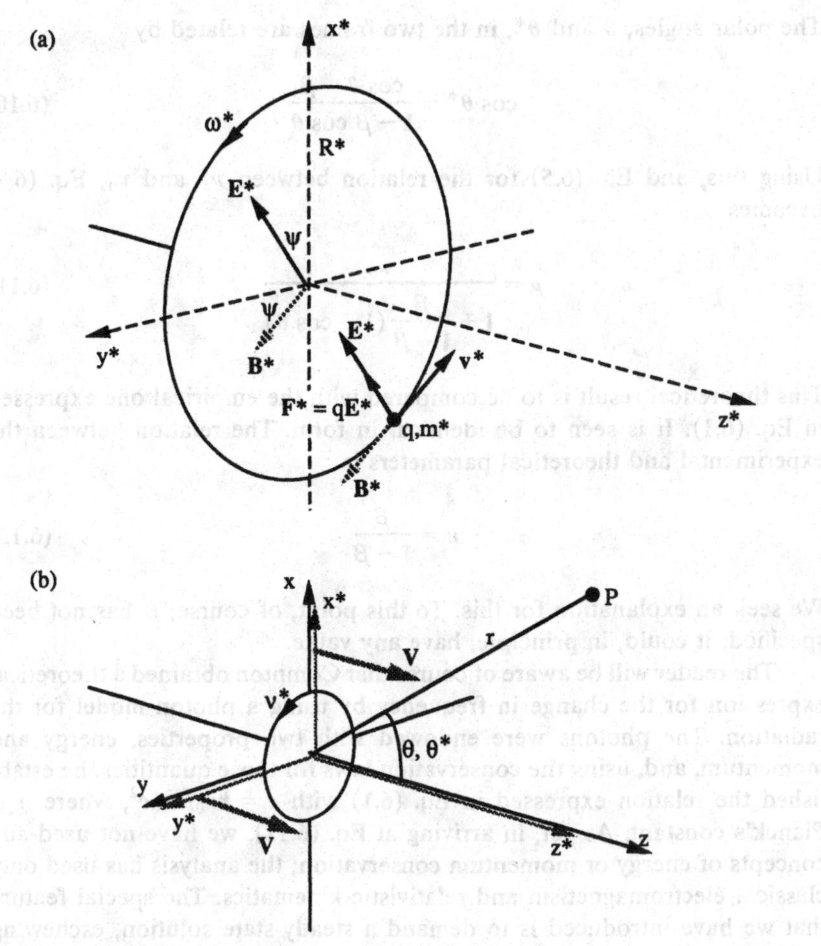

FIGURE 7. The Compton effect. (a) Motion of the charged particle (q, m_0) under the influence of the fields **E*** and **B*** in the zero-momentum (ZM) frame. (b) Emission of radiation at an angle θ^* in the ZM frame and at θ in the laboratory frame.

where m^* is the relativistic mass in the ZM frame, can be used to give the radius R^* of the steady-state orbit. This new steady state we shall call the *interaction state.*

The rotating charge radiates (or scatters) an electromagnetic wave which has a frequency ν^* in the ZM frame. According to the relativistic Doppler formula, the frequency observed in the laboratory (i.e., at the point P, at an angle θ in the L frame as shown in Fig. 7b) is

$$\nu = \nu^* \frac{1 + \beta \cos\theta}{(1 - \beta^2)^{1/2}} \tag{6.9}$$

The polar angles, θ and θ^*, in the two frames are related by

$$\cos\theta^* = \frac{\cos\theta - \beta}{1 - \beta\cos\theta} \tag{6.10}$$

Using this, and Eq. (6.5) for the relation between ν^* and ν_0, Eq. (6.9) becomes

$$\nu = \frac{\nu_0}{1 + \dfrac{\beta}{1-\beta}(1 - \cos\theta)} \tag{6.11}$$

This theoretical result is to be compared with the empirical one expressed in Eq. (6.1). It is seen to be identical in form. The relation between the experimental and theoretical parameters is

$$\varepsilon = \frac{\beta}{1-\beta} \tag{6.12}$$

We seek an explanation for this. To this point, of course, β has not been specified; it could, in principle, have any value.

The reader will be aware of course that Compton obtained a theoretical expression for the change in frequency by using a photon model for the radiation. The photons were endowed with two properties, energy and momentum, and, using the conservation laws for those quantities, he established the relation expressed in Eq. (6.1) with $\varepsilon = h\nu_0/m_0c^2$, where h is Planck's constant. As yet, in arriving at Eq. (6.11), we have not used any concepts of energy or momentum conservation; the analysis has used only classical electromagnetism and relativistic kinematics. The special feature that we have introduced is to demand a steady-state solution, eschewing completely any discussion of the transient behavior between the initial state and the interaction state. We now consider the energy and momentum of the system in order to establish the laws of interaction, and to establish a specific value for β.

6.3. THE LAWS OF INTERACTION BETWEEN RADIATION AND MATTER

In order to establish a connection between the factor $\beta/(1-\beta)$ obtained in Eq. (6.11) and physical constants, we must make use of the conservation laws. In the transition from the initial steady state to the interaction steady state, there is a change of linear momentum $\mathbf{P}$ and of total energy W of the charged particle; a quantity of momentum $\mathbf{p}$ and a quantity of energy w must have been extracted from the electromagnetic field to accomplish this change. The classical theory of radiation, based on Maxwell's equations, tells us that the energy per unit volume, $u = \varepsilon_0 E_0^2$, and the momentum per unit volume, $g = \varepsilon_0 E_0^2/c$, are related by $u = cg$. Hence, the amounts of energy and momentum extracted from the electromagnetic field to effect the transition between the two steady states are related by $w = cp = c|\mathbf{p}|$. Temporarily, until we link with more well-known terminology in physics, we shall refer to this as a "tantum" of the electromagnetic field, an amount of energy and momentum extracted from the field in order to effect a change in the energy and momentum of the particle with which it is interacting. The description of momentum and energy interchange between the initial steady state before interaction and the new steady state after interaction is illustrated in the top four boxes of Fig. 8. In the top row the particle motion is shown in the incident state (before interaction), and in the second row it is shown in the interaction state (after interaction); the left-hand column depicts the situation in the L frame, and the right-hand column depicts the situation in the ZM frame. Translation between horizontal neighboring boxes (the L and ZM frames) is achieved by using Lorentz transformation; vertical neighbors (the incident and interaction states) are connected by the conservation laws. The final state, after scattering, depicted in the bottom row, will be discussed later.

We start the discussion of momentum and energy in the top-left box where the incident state in the L frame is defined by the values ν_0 for the frequency of the radiation, $P_0 = 0$ for the momentum of the charged particle, and $W_0 = m_0c^2$ for the total energy of the particle. The corresponding values for ν^*, P_0^*, W_0^* in the ZM frame are established by the Lorentz transformation

$$\nu^* = \gamma(1 - \beta)\nu_0$$

$$P_0^* = -\gamma\beta m_0 c$$

$$W_0^* = \gamma m_0 c^2$$

In this zero-momentum frame we can immediately write $p_0^* = -P_0^*$ for the "tantum" of field momentum. This "tantum" has an energy $w_0^* = cp_0^*$. Then the values of p_0 and w_0, the "tantum" of momentum and energy extracted from the incident radiation in the L frame, can be obtained by reverse Lorentz transformation; these are

$$p_0 = \frac{\beta}{1 - \beta} m_0 c$$

$$w_0 = \frac{\beta}{1 - \beta} m_0 c^2$$

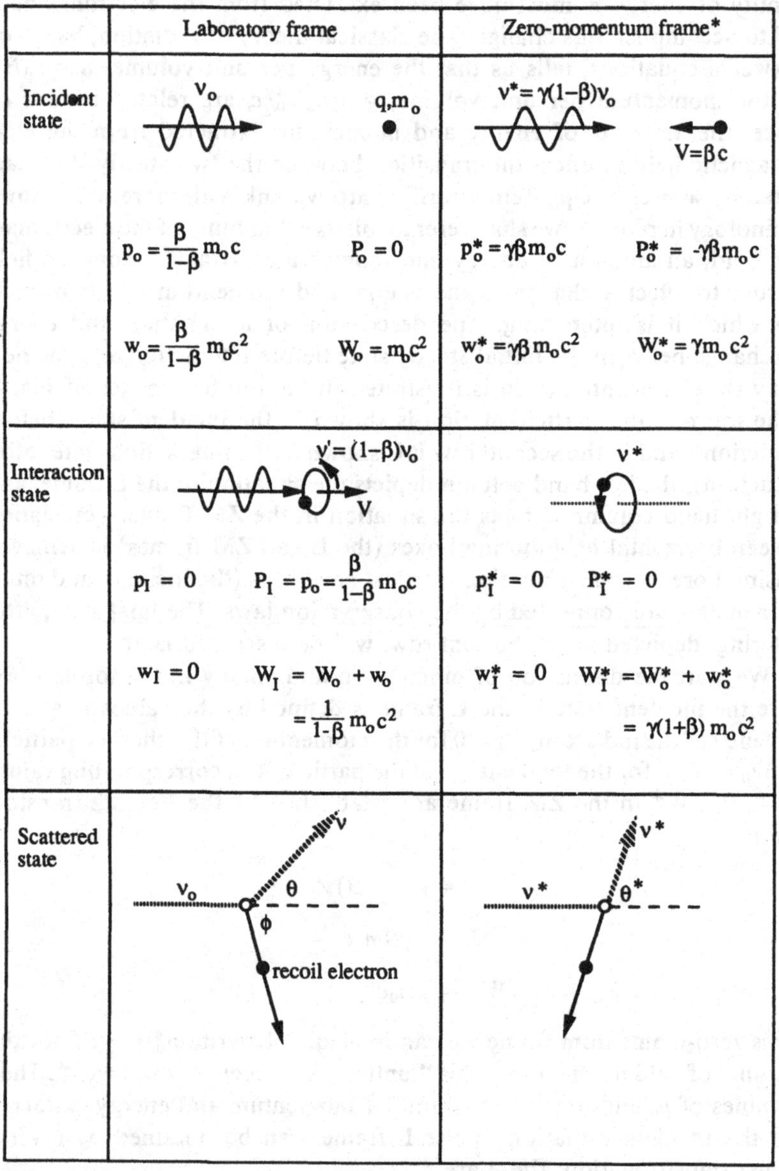

	Laboratory frame	Zero-momentum frame*
Incident state	ν_o q, m_o $p_o = \frac{\beta}{1-\beta} m_o c$ $P_o = 0$ $w_o = \frac{\beta}{1-\beta} m_o c^2$ $W_o = m_o c^2$	$\nu^* = \gamma(1-\beta)\nu_o$ $V = \beta c$ $p_o^* = \gamma\beta m_o c$ $P_o^* = -\gamma\beta m_o c$ $w_o^* = \gamma\beta m_o c^2$ $W_o^* = \gamma m_o c^2$
Interaction state	$\nu' = (1-\beta)\nu_o$ V $p_I = 0$ $P_I = p_o = \frac{\beta}{1-\beta} m_o c$ $w_I = 0$ $W_I = W_o + w_o = \frac{1}{1-\beta} m_o c^2$	ν^* $p_I^* = 0$ $P_I^* = 0$ $w_I^* = 0$ $W_I^* = W_o^* + w_o^* = \gamma(1+\beta) m_o c^2$
Scattered state	ν_o ν θ ϕ recoil electron	ν^* ν^* θ^*

FIGURE 8. The scattering of radiation by a free particle, of charge q and mass m. The three vertical levels of the diagram represent the states of the system before, during, and after scattering; they are related to one another by the conservation laws. The two columns represent the states of the system as observed in two frames of reference, laboratory and zero-momentum. P and W are the linear momentum and the total energy of the particle; p and w are the momentum and energy of that "tantum" of the field required to satisfy the conservation laws.

After interaction (important quantities in the interaction state are labeled by a subscript I), p_I^* and P_I^* are both zero. Therefore $w_I^* = 0$. Using the conservation of energy, the total energy of the now orbiting charge is

$$W_I^* = W_0^* + w_0^* = \left(\frac{1+\beta}{1-\beta}\right)^{1/2} m_0c^2$$

$$= \gamma(1+\beta)m_0c^2 \tag{6.13}$$

By reverse transformation, the values of P_I and W_I in the L frame are established. All values of momentum and energy for the charged particle and for the radiation involved are now established, before and after interaction, in both the L and ZM frames. The following changes of energy and momentum have occurred in the interaction. A "tantum of energy"

$$\Delta W = w_0 - w_I = W_I - W_0$$

$$= \frac{\beta}{1-\beta} m_0c^2 \tag{6.14}$$

has been transferred from the field to the particle. Using the identification of Eq. (6.12) and the empirical result of Eq. (6.2) we obtain

$$\Delta W = (\kappa c^2)\nu_0 \tag{6.15}$$

The same result is obtained in the ZM frame but with ν^* on the right-hand side.

Similarly, a "tantum" of linear momentum

$$\Delta P = p_0 - p_I = P_I - P_0$$

$$= \frac{\beta}{1-\beta} m_0c = (\kappa c^2)\frac{\nu_0}{c} = (\kappa c^2)\frac{1}{\lambda_0} \tag{6.16}$$

has been transferred from the field to the particle ($\lambda_0 = c/\nu_0$ is the wavelength of the radiation in the L frame). Again the same result is obtained in the ZM frame with $\lambda^* = c/\nu^*$ on the right-hand side.

Angular momentum also has been transferred to the particle. A formula for angular momentum, appropriate to the present situation and justified in Appendix 4, is

$$L = \frac{W^2 - c^2P^2 - m_0^2c^4}{4\pi\nu(W - cP)} \tag{6.17}$$

where, ν, P, and W are the driving frequency, the linear momentum, and the total energy of the particle in the frame being used. In the incident state for both the L and ZM frames, $W^2 - c^2P^2 = m_0^2c^4$, and the angular momentum charged particle is zero. For the interaction state, substitution of the appropriate values from Fig. 8 leads to $L = (\kappa c^2)/2\pi$ in both frames. A "tantum" of angular momentum:

$$\Delta L = \frac{(\kappa c^2)}{2\pi} \tag{6.18}$$

has been transferred from the field to the particle.

Since the physical constant κc^2 has already occurred in three equations it deserves a symbol of its own; using the value of κ given below Eq. (6.2) it is evaluated to be

$$\begin{aligned} h &= \kappa c^2 \\ &= (7.3726 \times 10^{-51} \text{ kg s}) \times (2.99793 \times 10^8 \text{ m s}^{-1})^2 \\ &= 6.6262 \times 10^{-34} \text{ J s} \end{aligned} \tag{6.19}$$

The "newly discovered" physical constant, h, is just Planck's constant, introduced by him to explain the distribution of radiation from a black body. We defer to his nomenclature and write the ***quantum** rules for the interaction of an electromagnetic wave with matter*:

> When electromagnetic radiation of frequency ν interacts with a charge-matter system, it causes
>
> - a change of *energy* of $h\nu$,
> - a change of *linear momentum* of $h\nu/c = h/\lambda$,
> - a change of *angular momentum* of $h/2\pi$,
>
> where Planck's constant $h = 6.6262 \times 10^{-34}$ J s.

The Compton effect has been analyzed by this rather unusual route for a number of reasons. First of all it establishes the same rule for energy change in interaction as did Planck, and therefore provides an alternative entry to quantum physics—history as it might have been. Secondly it establishes, at the same time, the rules for linear and angular momentum change in interaction, and therefore gives some prior justification for the postulates, made by Bohr in his theory of the hydrogen atom, and for the wave theory of matter, the precursor of quantum theory. Thirdly, from a didactic point of view, the approach has enabled the Compton effect to be discussed while maintaining contact with classical models. Such models are still useful as we enter the area of the quantum description of atoms. It is therefore appropriate to retain a classical description for this simplest of all interactions between an electromagnetic field and matter and to avoid the presumption, as is made in so many text books, that the Compton effect can only be treated by a quantum model of radiation. It may of course yet prove necessary to quantize the radiation, but, at the moment, we have achieved an agreement between experiment and theory by using a classical description. The consequence of requiring that the states of the system before, during, and after interaction shall be stationary states has led to the above stated *quantum rules for interaction.*

It will have been noted that, as yet, we have made no postulation of light existing as corpuscles or photons with energy and momentum determined by the frequency of the wave they replace. Nor has an intrinsic spin angular momentum of $h/2\pi$ been introduced. We have described the radiation by an electromagnetic wave, and have introduced certain quantum rules to govern the interaction process (some discussion of the different models will be undertaken in Chapter 14).

Having achieved the quantum rules for interaction we can immediately write the Compton scattering equation in its final form. From Eqs. (6.2), (6.12), and (6.19), we write

$$\varepsilon = \frac{\beta}{1-\beta} = \frac{\kappa \nu_0}{m_0} = \frac{h\nu_0}{m_0 c^2} \tag{6.20}$$

Substitute this result in Eq. (6.11):

$$\nu = \frac{\nu_0}{1 + \dfrac{h\nu_0}{m_0 c^2}(1 - \cos\theta)} \tag{6.21}$$

or, expressed in wavelength,

$$\lambda = \lambda_0 + \lambda_C(1 - \cos\theta) \tag{6.22}$$

The quantity $\lambda_C = h/m_0 c$ is the Compton wavelength earlier mentioned in Section 6.1.

6.4. THE SCATTERED STATE

The interchange of momentum and energy between the radiation and the charged particle, according to the rules we have established, is illustrated, for the ZM frame, in the upper two boxes of the right-hand column of Fig. 8. The transition between the interaction state and the scattered or final state must surely occur by the same rules, i.e., the absorption and the emission of radiation must follow the same rules. This is illustrated in the lower two boxes of the right-hand column of Fig. 8. In general, the scattered radiation described in the ZM frame is an elliptically polarized wave of frequency ν^* at an angle θ^*, and the recoil particle has a velocity V at an angle $\pi - \theta^*$. Lorentz transformation back to the L frame leads to Eq. (6.11), or to Eq. (6.21), for the frequency of the scattered radiation at angle θ in the laboratory, and to a recoil particle at an angle ϕ given by

$$\cos\phi = (1+\varepsilon)\tan\frac{\theta}{2} \tag{6.23}$$

with momentum and total energy as required to satisfy the conservation laws. One is led to the view that the observation of scattered radiation in direction θ is always accompanied, within the indeterminacies caused by the finite aperture of the measuring apparatus and the lack of precise knowledge of the position of the scattering electron, by a unique recoil momentum for the electron at angle ϕ. Or conversely that, if a recoil electron is detected in a certain direction, then the direction in which radiation has been scattered is determined. This model, which has strong similarities to the elastic collision of spheres, suggests that radiation consists of corpuscles called "photons," with some localization in space and time. The stage is set for discussion about the "real nature of light". The author takes the point of view that the only "real" things are the observation of measurable events in detectors. The *nature of light* itself is unknowable. The only important thing is to have a model, and an algebra to accompany it, which enables the results of the experiments to be predicted. From this point of view a *wave model of light* (with quantum rules for interaction) may prove just as successful overall as a *corpuscular model* (which must, in any case, incorporate wavelike properties in order to explain interference and diffraction phenomena). In the treatment in this book we use the former model, the so-called semiclassical theory. The matter is further discussed in Chapter 14.

CHAPTER 7

The Quantum Structure of the Atom

The earlier chapters have discussed the interaction between electromagnetic radiation and a *free* charged particle (electron). In the last chapter it was found that, when the classical description is compared with observations from experiment, it becomes necessary to introduce certain quantum rules. We must now turn to the interaction between electromagnetic radiation and an electron *bound* in an atom (or molecule or condensed matter). Presumably the same rules of interaction apply—the energy of the state of the atom is changed by $h\nu$, and the angular momentum of the state is changed by $h/2\pi$. The other rule, concerning the change of linear momentum, can be neglected in most circumstances. Of course a change of momentum does occur but, since the atom is considerably more massive than the electron, the observable effects are small. Mention will be made from time to time, however, of momentum change. The more noticeable changes are in those properties that are *intrinsic* to the internal structure of the atom—its internal state of energy and its internal state of angular momentum.

We set out in this chapter to describe the states in which an atom is permitted to exist and the rules governing transition between states. This quantum structure of the atom is described in many treatises and texts and the subject cannot be covered in a single chapter. We shall attempt only to present the main arguments and the results that will be of use in later chapters. For complete treatments the reader is referred to Condon and Shortley,[1] and to Cohen-Tannoudji, Diu, and Laloë.[2]

7.1. TRANSITIONS IN ATOMS

When the first quantum rule, concerning the frequency of the radiation and the energy change in the matter, is combined with the observation that only certain discrete frequencies of radiation interact with atoms (the *line* spectra of emission and absorption spectroscopy), one is led to the conclusion that there can be only certain discrete states of energy for the atom. This is in agreement with the empirical discoveries which culminated in the establishment of the *Ritz combination principle* (1908), which states that each atom can be characterized by a set of numbers (called *terms*); the wave numbers or frequencies of the observed spectral lines are always given by the difference between these terms. The differences do not give lines of the same strength or intensity, and, indeed, some differences do not seem to be allowed at all since the spectral line corresponding to that difference is not observed. Hence there is a further need to establish transition probabilities and selection rules. The principle has its modern equivalent in the following:

> The internal structure of the atom can exist in a discrete set of *states* (labeled $1, 2, 3, \ldots, i, j, k, \ldots, m, n, \ldots$) with discrete energies W_1, W_2, $W_3, \ldots, W_i, W_j, W_k, \ldots, W_m, W_n, \ldots$. The state of lowest energy W_1 is referred to as the *ground state* of the atom; the others are *excited states*.

A transition between two of these states can occur in interaction with radiation of frequency ν, which is close to the difference frequencies:

$$\nu_{mn} = \frac{W_m - W_n}{h} \tag{7.1}$$

This conforms with the quantum rule discovered in Chapter 6. This must be supplemented by *selection rules* which define which transitions $m \leftrightarrow n$ are allowed, what are the transition probabilities, and which are not allowed. There are several different types of transition:

(i) When an atom is bathed in radiation containing a component of frequency ν close to one of the above difference frequencies (ν_{ki}, say), it will induce an upward transition if the atom is initially in the lower state i. This is called (stimulated) *absorption* and an amount of energy $h\nu$ at the frequency ν is removed from the field to excite the atom. The most likely state for the atom to be in initially is the ground state $i = 1$, so the absorption spectrum consists of removing those frequencies from the incident field that represent excitation from the ground state *to those excited states that have allowed transitions from the ground.*

(ii) If the atom is initially in the higher-energy state k the radiation can induce a downward transition. This is called *stimulated emission* and an amount of energy $h\nu$ at frequency ν is emitted by the atom and added to the field.

These stimulated transitions are discussed in Chapter 9.

(iii) Even if there is no external field, an atom in an excited state will emit radiation spontaneously in making a transition to a state of lower energy (provided the transition is not forbidden by the selection rules). A *spontaneous emission* from state k to state i adds an amount of energy $h\nu_{ki}$ to the field (or vacuum) at frequency ν_{ki}. The reverse process, spontaneous absorption by transition from i to k, does not occur.

This one-way process, which injects an arrow of time into this area of physics, is discussed in Chapter 10.

(iv) Another type of transition, also stimulated by the presence of radiation, is that from a bound (usually ground) state to an *unbound* state. The excited electron becomes free of the atom leaving it as a positive ion.

This process, the *photoelectric effect*, is discussed in Chapter 12.

We are led to a description of the atom such that it can exist in any one of a number of discrete states—which we now specify as $\ldots|i\rangle, |j\rangle, |k\rangle, \ldots$, a notation that will be explained in Section 7.3—with discrete energies $\ldots W_i, W_j, W_k, \ldots$, respectively. While the atom rests in one of these states it does not emit or absorb energy; the allowed states are *stationary states.* Emission or absorption of energy takes place when the atom makes a transition from one allowed or stationary state to another.

An excited state is not completely "stationary," of course, because, by the spontaneous process, it can decay to a state of lower energy. This is a weak process (the decay period is long compared to the period of the radiation associated with the transition) and we shall take the attitude that it is caused by some weak perturbation process that is ever present. With this proviso the allowed states are completely stationary. An atom in an allowed or stationary state will remain (forever) in that state unless perturbed by some external influence (radiation, collisions, etc., or the ever-present decay perturbation).

We are left with the following questions: What is it that determines that a particular state with a particular energy is allowed? What are the selection rules that determine which transitions are allowed? What determines the probability of transition and hence the strength of the spectral lines? The answers clearly lie in the third quantum rule, regarding the angular momentum change in the transition.

7.2. ALLOWED STATES

The allowed states have been labeled $|1\rangle, \ldots, |i\rangle, |j\rangle, |k\rangle, \ldots$. Any two of these states between which a radiative transition is allowed must differ in angular momentum, according to the third quantum rule, by $h/2\pi$. Classically speaking, this angular momentum comes from the orbital motion of the electron about the nucleus. Since every state is accessible from any other by some sequence of transitions (otherwise it would never have been recognized as one of the allowed states), the angular momenta of all states must differ from each other in steps of an integer (including zero) multiples of $h/2\pi$. A transition between two states is allowed if they differ in angular momentum by $h/2\pi$; a transition is not allowed if they have the same angular momentum or if they differ by higher multiples of $h/2\pi$. (This rule strictly applies to electric dipole transitions only; it must be relaxed to account for the much weaker types of electromagnetic transition, electric quadrupole, magnetic dipole, electric octupole, etc. Such transitions will not be further discussed.)

It would be simple to label the allowed states with an angular momentum $nh/2\pi$, where n is an integer, and to formulate a rule whereby transitions are allowed only between two states whose n labels differ by unity. Such a simple view is wrong for two reasons. Firstly, angular momentum is a vector quantity. It is therefore possible to have a transition between two states of the same *magnitude* of angular momentum but of different orientation so that the change or vector difference is $h/2\pi$. It is necessary therefore to consider the vector nature of angular momentum and, in particular, the quantum mechanical restrictions that are placed on its specification. Secondly, it turns out that the electron has an intrinsic angular momentum called *spin*, with a magnitude of $\frac{1}{2}(h/2\pi)$. Thus the total angular momentum of an atom is a vector combination of the orbital angular momentum of the electron(s) and the spin angular momentum of the electron(s). To this must also be added a spin contribution from the nucleus—this is an integer (often zero) or half-integer* multiple of $h/2\pi$ depending on whether the mass number is even or odd. We shall ignore the refinement caused by nuclear spin which gives rise to hyperfine structure of spectral lines.

The combination of orbital and spin motion of the electron leads to the following properties for allowed atomic states:

- For atoms with an odd number of electrons (odd Z), the total angular momentum is represented by half-integer multiples of $h/2\pi$;
- For atoms with an even number of electrons (even Z), the total angular momentum is represented by integer multiples of $h/2\pi$;
- The angular momenta of the allowed states of a particular atom always differ by integer steps of $h/2\pi$.

The way in which allowed states of the atom are labeled to specify the angular momentum is discussed in Section 7.4. Each allowed state has some unique specification involving the angular momenta of the individual electrons and of the atom as a whole. Each allowed state has a particular energy—called, as we shall see below, the *eigenvalue* (or allowed value) of the energy. How to find this energy from the specification of the state is a

* "Half-integer" means "half of an odd integer, such as $\frac{1}{2}, \frac{3}{2}, \frac{5}{2}, \ldots$ etc."

difficult matter—strictly, it can only be solved algebraically for the hydrogenic atom (nuclear charge Ze, with one orbiting electron of charge $-e$; for hydrogen $Z = 1$).

Bohr constructed a solution using the purely classical model of a central heavy nucleus (charge Ze) with an orbiting electron (charge $-e$) held by the centripetal force of the Coulomb interaction. One can readily show that, for an orbital radius of r, the total energy (kinetic plus potential) of the system is

$$W = -\frac{\mu_0}{4\pi}\frac{Ze^2c^2}{2r} \tag{7.2}$$

Using the expression for the orbital angular momentum

$$L = \left(\frac{\mu_0}{4\pi} Ze^2c^2mr\right)^{1/2} \tag{7.3}$$

Equation (7.2) can be expressed as

$$W = -\left(\frac{\mu_0}{4\pi}\right)^2 \frac{Z^2e^4c^4m}{2}\frac{1}{L^2} \tag{7.4}$$

Bohr then placed the restriction that L should be an integer multiple ($n = 1, 2, 3, \ldots, \infty$) of $h/2\pi$ (note that in classical theory the orbital angular momentum cannot be zero; this restriction is not retained in quantum theory). The energy of this state is then

$$W_n = -Z^2\frac{\mu_0^2e^4c^4m}{8h^2}\frac{1}{n^2} = -Z^2hcR_\infty\frac{1}{n^2} \tag{7.5}$$

where $R_\infty = \mu_0^2e^4c^3m/8h^3 = 1.097\ldots \times 10^7\ \mathrm{m}^{-1}$ is the Rydberg constant. The factor hc converts the Rydberg, which is in units of wave number, the traditional unit of the spectroscopist, into units of energy. The equation is expressed in this way since, for many years, the Rydberg was one of the most accurately known of the physical constants.

When this is substituted into Eq. (7.1) the spectral frequencies (or wave numbers) of the hydrogen line spectrum that are observed in experiment are very accurately predicted—a most striking success.

Furthermore, through Eq. (7.2) or (7.3), one can predict the radius of the orbits:

$$r_n = \left(\frac{\mu_0}{4\pi}\right)^{-1}\frac{h^2}{4\pi^2Ze^2c^2m}n^2 = a_B\frac{n^2}{Z} \tag{7.6}$$

where

$$a_B = \left(\frac{\mu_0}{4\pi}\right)^{-1} \frac{h^2}{4\pi^2 e^2 c^2 m} = 5.292\ldots \times 10^{-11}\ \text{m} \tag{7.7}$$

is the Bohr radius, the radius of the first ($n = 1$) allowed orbit in the Bohr model of hydrogen ($Z = 1$). Although this concept of "radius of orbit" is not retained in the quantum-mechanical description of the atom, it still has a meaning of "some sort of mean radius for the atom" and a_B, being a physical constant, provides an appropriate scale for atomic dimensions. We may note in passing that $a_B = r_0/\alpha^2$, where r_0 is the classical radius of the electron [see Eq. (3.22), $r_0 = (\mu_0/4\pi)(e^2/m) = 2.818\ldots \times 10^{-15}$ m] and

$$\alpha = \frac{\mu_0}{4\pi}\frac{2\pi e^2 c}{h} = \frac{1}{137.036\ldots}$$

is the dimensionless fine-structure constant. Thus the Bohr radius is some 18800 times larger than the classical electron radius, a figure that gives an idea of the comparative emptiness of the atom.

This classical concept of the electron as a point particle in an orbit cannot be sustained, however—in particular it is found that many electronic states have zero angular momentum, which is not possible in the classical description. One must move to a quantum-mechanical description where, rather than an electron orbit, an electron distribution is defined by a probability wave function (to be further mentioned in Section 7.10). It is not our intention here to develop this description logically or fully. Here we shall confine our attention to the manner in which atomic states are specified and we shall quote a few results that will be needed in the subsequent treatment.

7.3. STATE VECTORS

Each allowed state of an atom is specified by a set of labels representing the values of certain physical attributes of the atom when it is in that state. The number of labels must be sufficient that the specification of each state is unique; each allowed state has its own unique set of labels. An allowed state is unique in the sense that it cannot be thought of as a mixture or superposition of other allowed states. The set of allowed states has, therefore, a similar aspect to a set of orthogonal vectors such as the unit vectors **i**, **j**, **k** of three-dimensional space. We shall therefore adopt a notation that suggests this "vector" character for the allowed states.

The vector representing the allowed state with labels α, β, γ (we shall, for the moment, assume three to be necessary, although more may be required) is written as $|\alpha\beta\gamma\rangle$. With it goes a conjugate vector written as $\langle\alpha\beta\gamma|$, in a similar manner in which an ordinary 3-space vector with complex coefficients $\mathbf{A} = \mathbf{i}a + \mathbf{j}b + \mathbf{k}c$ has a conjugate $\mathbf{A}^* = \mathbf{i}a^* + \mathbf{j}b^* + \mathbf{k}c^*$ such that the inner product $\mathbf{A}^* \cdot \mathbf{A} = a^*a + b^*b + c^*c$ is a positive quantity. The inner product of the state vector is written as $\langle\alpha\,\beta\,\gamma|\alpha\,\beta\,\gamma\rangle$. The vector $\langle\alpha\,\beta\,\gamma|$ is called a *bra*-vector and the vector $|\alpha\,\beta\,\gamma\rangle$ is called a *ket*-vector so that the inner product $\langle\alpha\,\beta\,\gamma|\alpha\,\beta\,\gamma\rangle$ closes the bra(c)ket and the result is a positive scalar quantity.

These state vectors representing the allowed states of the atom are not, of course, vectors in ordinary 3-space like **i**, **j**, **k**. The labels α, β, γ, . . . , each specifying the values of some physical attributes of the state, may take on a range of (sometimes infinite) values so that the vectors may span a space of many (sometimes infinite) dimensions. The space is called Hilbert space and any general state of the atom may be expanded in terms of the chosen set of allowed states.

From what has been said above, if any one of the labels is changed to a different (allowed) value, a different allowed state is produced. As with ordinary 3-space vectors, the inner product of a bra with a different ket would give zero; the vectors are therefore orthogonal. Furthermore we can specify that the vectors be unit vectors. Thus

$$\langle\alpha\,\beta\,\gamma|\alpha'\,\beta'\,\gamma'\rangle = \delta(\alpha\alpha')\delta(\beta\beta')\delta(\gamma\gamma') \tag{7.8}$$

the allowed vectors are *orthonormal.*

From the ordinary 3-space unit vectors dyadic tensor products can be formed with 9 elements **ii ij ik ji**, . . . , the trace of which is $\mathbf{ii} + \mathbf{jj} + \mathbf{kk} = \mathbf{I}$, the idem-tensor. **I** has the property that, in dot-product with any vector $\mathbf{A} = \mathbf{i}a_i + \mathbf{j}a_j + \mathbf{k}a_k$, it yields the same vector: $\mathbf{A} \cdot \mathbf{I} = \mathbf{I} \cdot \mathbf{A} = \mathbf{A}$. An idem-tensor appropriate to the Hilbert space can also be constructed:

$$\mathbf{I} = \sum_j |j\rangle\langle j| = \sum_{\text{all}} |\alpha\,\beta\,\gamma\rangle\langle\alpha\,\beta\,\gamma| = \sum_{\text{all}} |\alpha\,J\,M\rangle\langle\alpha\,J\,M| \tag{7.9}$$

the sum being over all the unit vectors that span the space (the final expression anticipates the labeling established below).

Any general vector of the system can be expanded in terms of the complete set of allowed vectors:

$$|\ \rangle = \cdots c_i|i\rangle + c_j|j\rangle + c_k|k\rangle \cdots \tag{7.10}$$

the c_j being called the amplitudes (or probability amplitudes). It is usual to define each state vector as a unit vector. Hence we have

$$\langle\ |\ \rangle = \cdots c_i^*c_i + c_j^*c_j + c_k^*c_k \cdots = 1$$

The modulus squared of the (probability) amplitude is the probability of the atom being found in the corresponding allowed state. It is readily seen, by multiplying each side of Eq. (7.10) by $\langle j|$, that each amplitude can be expressed as

$$c_j = \langle j|\ \rangle \tag{7.11}$$

Consequentially the general vector can be written as

$$\begin{aligned} |\ \rangle &= \cdots |i\rangle\langle i|\ \rangle + |j\rangle\langle j|\ \rangle + |k\rangle\langle k|\ \rangle \cdots \\ &= \sum_j |j\rangle\langle j|\ \rangle = \mathbf{I}|\ \rangle \end{aligned} \tag{7.12}$$

An allowed state of the atom has a specific value for its energy. In the language of quantum mechanics, an allowed state of an atom is an eigenstate of the Hamiltonian operator with an eigenvalue that is the allowed energy. Expressed in mathematical terms we have

$$\hat{H}_A|\alpha\,\beta\,\gamma\rangle = W_{\alpha\beta\gamma}|\alpha\,\beta\,\gamma\rangle \tag{7.13}$$

which can be read as "the result of operating on an allowed state of the atom with the Hamiltonian operator, which represents the intrinsic energy of the atom, yields the allowed energy of that state of the atom." The symbol $\hat{H}_A$ represents the Hamiltonian operator. (The caret is used to indicate quantities that are quantum-mechanical operators, and the subscript A indicates that the Hamiltonian represents the intrinsic or internal energy of the atom. This includes the Coulomb potential energies between the charged constituents, the kinetic energy of internal motion of the constituents and the magnetic interaction between the currents created by the charged constituents; it does not include any kinetic energy of motion of the atom as a whole, nor any potential energy due to any interaction with external influences). The symbol $W_{\alpha\beta\gamma}$ is the energy of the state $|\alpha\,\beta\,\gamma\rangle$ and the subscripts $\alpha\beta\gamma$ on W indicate that this will depend on the values of the labels that define the state.

Thus, in mathematical language, the allowed states of the atom, which are also stationary states, are *eigenstates* of the Hamiltonian operator which describes the intrinsic energy of the system. And the allowed values of the energy are the *eigenvalues* of the Hamiltonian.

We must now find a set of appropriate labels $\alpha, \beta, \gamma, \ldots$ for the vectors representing the allowed states of the atom, states which, apart from spontaneous decay, are stationary. The labels will represent a set of physical quantities whose values, once specified for the state, will give a sufficient and unique description of that allowed state of the atom.

7.4. LABELING THE ALLOWED STATE (EIGENSTATE) VECTORS

From the third quantum rule, that stating that in an atomic transition the angular momentum must change by $h/2\pi$, we are immediately led to the conclusion that allowed states must have unique values of angular momenta, and that they must differ by steps of $h/2\pi$. Being a vector, the angular momentum $\mathbf{J}$ has three components J_x, J_y, J_z with respect to some chosen coordinate system. However, in quantum mechanics, these three components cannot all be simultaneously known—they are incompatible observables. What can be known is

- The *magnitude* of the angular momentum, either as $|\mathbf{J}|$, or as its square

$$\mathbf{J}^2 = \mathbf{J} \cdot \mathbf{J};$$

- The magnitude of *one* of its components, usually chosen as J_z.

It is shown in texts on quantum mechanics (see for example Cohen-Tannoudji, Diu, and Laloë,[(2)] Chapter 6) that $\mathbf{J}^2$ can only have the values $J(J+1)(h/2\pi)^2$, where J is either an integer $0, 1, 2, 3, \ldots$ (for atoms with an even number of electrons), or a half-integer $\frac{1}{2}, \frac{3}{2}, \frac{5}{2}, \ldots$ (for atoms with an odd number of electrons). As mentioned earlier, we neglect the influence of nuclear spin.

And the component J_z can only have the values $Mh/2\pi$ where, for any given value of J, M can take on the values in integer steps from $-J$ to $+J$. (The label M is sometimes written as M_J to indicate that it belongs to the J quantum number; this may be important in situations where it is necessary to distinguish the contributions to the total angular momentum made by various sources, orbital and spin motion.)

The two labels J and M can therefore be used to specify the total angular momentum state of the atom as far as it can be under the rules of quantum mechanics. Thus the eigenstates can be written as

$$|\alpha J M\rangle$$

where α represents a set of labels, as yet unspecified, that is required to complete a unique description of the eigenstate of the atom. This state has the values of

$$J(J+1)\left(\frac{h}{2\pi}\right)^2 \qquad \text{for the square of the angular momentum}$$

or (7.14)

$$[J(J+1)]^{1/2}(h/2\pi) \qquad \text{for the magnitude of the angular momentum}$$

and

$$M\frac{h}{2\pi} \qquad \text{for the } z \text{ component of the angular momentum} \tag{7.15}$$

Just as $W_{\alpha JM}$ is the eigenvalue of the Hamiltonian operator for the state $|\alpha\, J M\rangle$, so the above values are the eigenvalues of two further quantum operators representing the square of the angular momentum and its z component. Thus

$$\hat{\mathbf{J}}^2|\alpha\, J M\rangle = J(J+1)\left(\frac{h}{2\pi}\right)^2|\alpha\, J M\rangle \tag{7.16}$$

$$\hat{J}_z|\alpha\, J M\rangle = M\frac{h}{2\pi}|\alpha\, J M\rangle \tag{7.17}$$

Note that the operators $\hat{\mathbf{J}}$ and $\hat{J}_z$ have dimensions of angular momentum, whereas the labels J and M are dimensionless numbers. The number J is referred to as the *angular momentum quantum number* because, through Eq. (7.14), it defines the allowed value of the magnitude of the angular momentum. The number M is often referred to as the *magnetic quantum number.* States with the same values of α and J are normally degenerate in M, i.e., all states of a given α and J with the various possible values of M have the same energy, as we would expect from the properties of isotropic space where the energy of an atom cannot depend on its orientation. The energy-eigenvalue equation can be written as

$$\hat{H}_A|\alpha\, J M\rangle = W_{\alpha J}|\alpha\, J M\rangle \tag{7.18}$$

the energy depending only on the values of the labels α and J. But this degeneracy is removed when a magnetic field is established, removing the isotropy of space. This shift of energy levels according to their M values when a magnetic field is applied is called the Zeeman effect [see Eq. (7.21) below and in Section 7.8]. A somewhat similar effect is caused by an applied electric field—the Stark effect.

According to Eqs. (7.16), (7.17), and (7.18) the allowed states $|\alpha\, J M\rangle$ are, simultaneously, eigenstates of the quantum-mechanical operators $\hat{H}_A$, $\hat{\mathbf{J}}^2$, and $\hat{J}_z$. These three operators are called *commuting operators.* Two quantum operators $\hat{A}$ and $\hat{B}$ are commuting operators when the commutator

$[\hat{A}, \hat{B}] = \hat{A}\hat{B} - \hat{B}\hat{A} = 0$. It can readily be shown that if $\hat{A}$ and $\hat{B}$ are any pair of the above atomic operators then, when $\hat{A}\hat{B} - \hat{B}\hat{A}$ is applied to the state vector $|\alpha\, J\, M\rangle$ the result is zero. There may be other operators that also commute with the Hamiltonian. One such is the Hamiltonian operator representing the energy of interaction between the atom and an applied magnetic field. The classical expression for the energy of interaction between a magnetic dipole $\mathbf{m}$ and a magnetic field $\mathbf{B}$ is $-\mathbf{m} \cdot \mathbf{B}$. The quantum mechanical expression for the energy operator is

$$\hat{H}_{\text{mag}} = -\hat{\mathbf{m}} \cdot \mathbf{B} \tag{7.19}$$

The operator corresponding to the magnetic moment is given by

$$\hat{\mathbf{m}} = -\frac{2\pi g \hat{\mathbf{J}} \mu_B}{h} \tag{7.20}$$

where $\mu_B = eh/4\pi m_e = 9.274 \times 10^{-24}$ A m^2 is the Bohr magneton, the atomic unit of magnetic moment, and g is the Landé factor. If the direction of $\mathbf{B}$ is chosen as the direction of the z axis, Eq. (7.19) becomes

$$\hat{H}_{\text{mag}} = \frac{2\pi g \mu_B B}{h} \hat{J}_z \tag{7.21}$$

Because $\hat{J}_z$ commutes with $\hat{H}_A$, $\hat{H}_{\text{mag}}$ also commutes with $\hat{H}_A$. $\hat{H}_{\text{mag}}$ has the eigenvalue equation

$$\hat{H}_{\text{mag}} |\alpha\, J\, M\rangle = g\mu_B B M |\alpha\, J\, M\rangle$$

There are other contributions to the total Hamiltonian of the system (i.e., representing all contributions to the energy of the atom, internal and external) which do not commute with $\hat{H}_A$. Two such that will be met with in later chapters are $\hat{H}_D$, describing the process that leads to the spontaneous decay of an excited state and therefore to radiation of energy and energy reduction of the atom; and $\hat{H}_{\text{int}}$, describing the interaction between the electric dipole moment of the atom, $\hat{\mathbf{D}} = -e \sum \hat{\mathbf{r}}_i$ (we use $\mathbf{r}$ for the displacement of an electron in an atom from some origin in the atom; the sum is over all electrons in the atom), and an applied electric field $\mathbf{E}$: $\hat{H}_{\text{int}} = -\hat{\mathbf{D}} \cdot \mathbf{E}$. This is responsible for the Stark effect mentioned above and for the very important electric dipole radiation associated with the transition between certain states in the atom.

7.5. THE CONFIGURATION

We have introduced labels J and M to specify what can be known about the final state of angular momentum for the atom. We must now enquire what other information is necessary to complete the unique specification of each state. This information has been summarized in the label α. This is called the configuration, a specification of the state of each electron in the atom, and how the angular momenta of the individual electrons are coupled together to produce the final state specified by J and M. We start by considering the situation for the hydrogenic atom. The quantum-mechanical description of this can be completely specified but will not be given here—the reader must seek that in books on the quantum structure of the atom.

The state of the atom with one electron (or, what is equivalent, the state of the electron in a one-electron atom) requires four labels—effectively three because the electron has three degrees of freedom in space and an extra one to specify the orientation of its spin. The four labels are as follows

- n, the principal quantum number. This specifies the radial distribution and is closely related to the n introduced in the Bohr theory in Eq. (7.5). As it does for the classical radius in Eq. (7.6), it determines, in a loose way, the radial size of the wave function describing the probability distribution of the electron in space. It takes on the integer values $n = 1, 2, 3, \ldots, \infty$ with $n = 1$ being the closest bound and lowest energy level.
- l, the orbital quantum number. This specifies the magnitude squared of the *orbital* angular momentum of the electron, $\mathbf{L}^2 = l(l+1)(h/2\pi)^2$. For any values of n, l can take on the integer values $l = 0, 1, 2, \ldots, \leq n-1$ (n possible values). Note that, in the quantum description of the atom, the orbital angular momentum can be zero ($l = 0$).
- m_l, the orbital projection quantum number. This specifies the projection of the orbital angular momentum $\mathbf{L}$ along the z axis, $L_z = m_l h/2\pi$. For any value of l, m_l can take on the integer values $m_l = -l$, $-l+1$, $-l+2, \ldots, 0, \ldots, l-2$, $l-1$, l, ($2l+1$ possible values).
- m_s, the spin projection quantum number. The electron has an intrinsic spin angular momentum whose magnitude is given by $|\mathbf{S}| = [s(s+1)]^{1/2}(h/2\pi)$, with $s = \frac{1}{2}$, so that $|\mathbf{S}| = (\sqrt{3}/2)(h/2\pi)$ and is the same for every electron. The number m_s specifies the projection of the spin angular momentum along the z axis, $S_z = m_s h/2\pi$. Independently of all other values and restrictions, m_s can take on two values, $-\frac{1}{2}$ and $+\frac{1}{2}$.

Every electron in the atom can be labeled by the four quantum numbers n, l, m_l, m_s. The values of these labels are restricted by a very important principle, *the Pauli Exclusion Principle: No two electrons in an atom can have the same set of four quantum numbers.* Stated first as an empirical principle, this explains much about the structure of atoms, and especially the shell structure of the periodic table. The principle is, however, a consequence of a requirement in quantum mechanics—that, for a system built from fermions (i.e., particles of spin given by $s = \frac{1}{2}$, as are electrons, protons, and neutrons), the wave function of the system must be antisymmetric.

The set of labels for all electrons in the atom is called the *configuration*, and is summarized by the label α in the notation $|\alpha\, J\, M\rangle$. The specification of the configuration is not as complicated as it might appear, i.e., a set of number labels for every electron of the atom. As one fills up the possible states, starting from the lowest with $n = 1$, and using the restrictions imposed by the limitations on the values for l and m_l, together with those imposed by the Pauli exclusion principle, *closed shells* are formed in which every possible orientation of orbital and spin angular momentum is occupied. The result is that, for these closed structures,

- the resultant orbital angular momentum for the shell, $\mathbf{L} = \mathbf{l}_1 + \mathbf{l}_2 + \mathbf{l}_3 \ldots = 0$,
- the resultant spin, $\mathbf{S} = \mathbf{s}_1 + \mathbf{s}_2 + \mathbf{s}_3 \ldots = 0$,
- the total angular momentum of the closed shell, $\mathbf{J} = \mathbf{L} + \mathbf{S} = 0$.

The angular momentum state of the atom as a whole is therefore determined by only those electrons outside closed shells. So, in practice, only the list of (nl) values of electrons outside the closed shells need be given. Furthermore, when an atom is excited, it is normally only one of the extracore electrons that is lifted out of its normal (nl) state so that the excited states are determined by how the orbit and spin of that excited electron is coupled to the orbital and spin of the remaining extracore electrons.

Rather than proceed with a general discussion it will suffice for our purpose to illustrate the complete specification of the states of an atom by considering two examples.

The Allowed States of Sodium

Sodium has eleven electrons. For the first level of $n = 1$, there is only one value of $l = 0$, and hence only one value of $m_l = 0$. However, there are two possible values for $m_s = \pm\frac{1}{2}$. Therefore two electrons fill the first shell $(1s)^2$. (Here we have introduced the private language of the spectroscopist: a spectroscopist does not specify the orbital by an integer for l, but calls $l = 0$ an s-orbital, $l = 1$ a p-orbital, $l = 2$ a d-orbital, and $l = 3$ an f-orbital; this apparently random set of letters is followed by the alphabetical sequence $g, h, i, \ldots$ for higher values of $l = 4, 5, 6, \ldots$.) Having now filled the $n = 1$ level, the next electrons must go into the $n = 2$ level. The first two of these are in the $(2s)^2$ level, and the next six go into $(2p)^6$—this shell of electrons can take six electrons because, with $l = 1$, there are three possible values of $m_l = -1, 0, +1$ each with two values of $m_s = \pm\frac{1}{2}$. Thus the first ten electrons have filled up three shells—$(1s)^2(2s)^2(2p)^6$. This core contains all possible orientations for orbits and spins so it has zero angular momentum. Provided the core remains intact, the properties of the atom in its ground state and in its excited states are determined then by the single remaining extra-core electron—sodium is a typical *one-electron atom*, and its structure has some similarities to all other one-electron atoms, including hydrogen.

The ground state of the sodium atom has the single extracore electron in the $(3s)$ state (the next available with $n = 3$ and $l = 0$).

The orbital angular momentum of the atom (defined by the quantum number L) is due to the orbital angular momentum of that single electron ($l = 0$); therefore $L = l = 0$. Using spectroscopists' notation, this is referred to as an S-state (meaning $L = 0$; upper-case letters are used for the atomic state whereas lower-case letters are used for individual electrons).

The spin angular momentum of the atom (defined by the quantum number S) is due to the spin angular momentum of that single electron ($s = \frac{1}{2}$); therefore $S = s = \frac{1}{2}$. A state with $S = \frac{1}{2}$ is normally referred to as a "doublet" since there are two possible orientations for the spin relative to the orbital vector (in general, the projection of the **S** vector, specified by M_s, can take on the $2S + 1$ values from $-S$ in integer steps to $+S$; the quantity $2S + 1$ is called the *multiplicity* of the state. For $S = \frac{1}{2}$, the multiplicity is 2, a doublet state). In this case, however, $L = 0$, the orbit and spin can combine to form only one possible value for the total angular momentum $\mathbf{J} = \mathbf{L} + \mathbf{S}$; $L + S = 0 + \frac{1}{2} = \frac{1}{2} = J$.

Rules for coupling angular momenta. When two angular momenta such as **L**, specified by the quantum number L, and **S**, specified by the quantum number S, are coupled together to give the vector sum $\mathbf{J} = \mathbf{L} + \mathbf{S}$, what are the possible values for the quantum number J used to specify **J**? J can take on all values from the maximum $L + S$, in integer steps down to the minimum of $|L - S|$. Note that L, S, and J are always positive numbers or zero.

The ground state of sodium has, by this rule and in spite of the fact that its multiplicity is 2, only the one value of $J = 1/2$, $[|\mathbf{J}| = (\frac{1}{2} \cdot \frac{3}{2})^{1/2}(h/2\pi)]$, with two possible values for $M = \pm\frac{1}{2}$. As described above, the two states with different values for M are degenerate, but will become separated in the presence of a magnetic field. The final specification of this state is expressed by the ***state symbol*** $^2S_{1/2}$.

The ***state symbol*** is the spectroscopists' way of specifying L, S, and J for the state: $^{2S+1}L_J$, with one of the letters $S, P, D, F, \ldots$ in place of a numerical value for L. Note the two different uses of the letter S: Firstly for the spin quantum number of the atom; in the example just discussed $S = \frac{1}{2}$; secondly as a letter code standing for a state of zero angular momentum, $L = 0$.

The state vector for the ground state of sodium is written as $|3s\ J = \frac{1}{2}\ M = \pm\frac{1}{2}\rangle$ or $|3s\ \frac{1}{2}\ \pm\frac{1}{2}\rangle$.

When the sodium atom is excited the extracore electron is raised to a state of higher energy. The first three configuration states that are available are $(3p)$, $(3d)$, and $(4s)$. We shall discuss only the first of these, $n = 3$, $l = 1$. The orbital state of the atom is just that of the single electron, $L = 1$; the spin state of the atom is just that of the single electron, $S = \frac{1}{2}$. These couple together, by the rule given above, to form two possible values for the total angular momentum quantum number, $J = L + S$ and $L - S = \frac{3}{2}$ and $\frac{1}{2}$ in this case. Thus the configuration $(3p)$ leads to two possible atomic states: $|3p\ J = \frac{1}{2}\ M\rangle$ with two magnetic sublevels $M = \pm\frac{1}{2}$, and $|3p\ J = \frac{3}{2}\ M\rangle$ with four possible magnetic sublevels $M = \frac{3}{2}, \frac{1}{2}, -\frac{1}{2}, -\frac{3}{2}$. These states have the spectroscopic state symbols $(3p)\,^2P_{1/2}$, and $(3p)\,^2P_{3/2}$, respectively. These are the first two excited states of sodium both belonging to the $(3p)$ configuration. Radiative transitions from them to the ground state $(3s)\,^2S_{1/2}$ give the well-known D_1 and D_2 yellow lines of the sodium spectrum (with wavelengths of 589.6 and 589.0 nm, respectively).

The Allowed States of Mercury

This is a typical two-electron atom, i.e., with two electrons outside closed shells. Mercury has 80 electrons, the first 78 of which form a core with configuration $(1s)^2$ © $(2s)^2(2p)^6$ © $(3s)^2(3p)^6$ © $(3d)^{10}(4s)^2(4p)^6$ © $(4d)^{10}(5s)^2(5p)^6$ © $(4f)^{14}(5d)^{10}$, in which the symbols © indicate, respectively, the closed cores of the atoms of helium, neon, argon, krypton, and xenon, the inert gases. Outside these closed shells of the inert gases, are two further shells $(4f)^{14}(5d)^{10}$ that are both filled. The last two electrons of mercury go into $(6s)^2$ for the ground state. The two electrons both have orbital values $l = 0$; therefore the atomic orbital number is $L = 0$. The two electrons both have spins of $s = 1/2$ so, in general, they could combine with antiparallel spins to form $S = \frac{1}{2} - \frac{1}{2} = 0$, or with parallel spins to give $S = \frac{1}{2} + \frac{1}{2} = 1$. However, the latter is not allowed by the Pauli exclusion principle. The ground state of mercury has $L = 0$, $S = 0$ and therefore, $J = 0$ and $M = 0$. It is, indeed, like a closed subshell itself. The ground state vector is written $|6s^2\,0\,0\rangle$, and the state symbol is $6s^2\,^1S_0$. The state symbol is to be translated as follows: there are two electrons outside closed shells both in the $6s$ orbital; the total spin of the atom is $S = 0$, indicated by the "multiplicity" $= 2S + 1 =$ "singlet" $= 1$; the total orbital quantum number is $L = 0$, indicated by the letter code S; the total angular momentum is $J = 0$, indicated by the final subscript.

When the atom of mercury is excited, it is general that only one electron is lifted out of its $6s$ orbital, giving excited configurations of $6s6p$, $6s7s$, $6s7p$, etc. We shall briefly discuss only the first of these. In the $6s6p$ configuration the value of $L = l_1 + l_2 = 0 + 1 = 1$, one value only. However, for the spin, the restriction imposed by the Pauli exclusion principle has been removed (the electrons being now in different l states) so that there are two possible states with $S = 0$ (a "singlet" state) or with $S = 1$ (a "triplet" state).

For the *singlet* state $L = 1$ and $S = 0$, so there is only one value of $J = 1$. The state vector is $|6s6p\ 1\ M\rangle$ (there being three possible values of $M = -1, 0, +1$). And the state symbol is $6s6p\ ^1P_1$. Sometimes the state vector may be written as $|6s6p, (L\ S), J\ M\rangle = |6s6p, (1\ 0), 1\ M\rangle$, where the $(L\ S)$ gives an intermediate step after the statement of the configuration, showing how the individual electrons have coupled together to form the orbital and spin of the atom, before they, in turn, have combined together to form the final state of total angular momentum given by J. Care must be exercised, however, because the manner in which individual angular momenta couple together to produce the final result is not necessarily unique. For this reason they are placed in parentheses indicating that the L and S value may not be "good quantum numbers"; in other words the state of the atom may not be an eigenstate of the orbital operator $\hat{\mathbf{L}}^2$ or of the spin operator $\hat{\mathbf{S}}^2$, although it is always an eigenstate of $\hat{\mathbf{J}}^2$ and of $\hat{J}_z$. The matter will be briefly mentioned below in its relevance to mercury.

For the *triplet* state, $S = 1$ leads to three possible values of $J = L + S$, $L + S - 1$, $L - S = 2, 1, 0$ in this case. The "triplet" states are three in number:

- $|6s6p, (1\ 1), 2\ M\rangle$, with five possible values of $M = -2, -1, 0, +1, +2$; $6s6p\ ^3P_2$;
- $|6s6p, (1\ 1), 1\ M\rangle$, with three possible values of $M = -1, 0, +1$; $6s6p\ ^3P_1$;
- $|6s6p, (1\ 1), 0\ 0\rangle$, with only one value for $M = 0$; $6s6p\ ^3P_0$.

It will be noted that we now have a unique set of labels for the four eigenstates of the atom in its first excited state of configuration $6s6p$. There is one state each with $J = 0$ and $J = 2$. There are two states belonging to the $6s6p$ configuration that both have $J = 1$. They are distinguished by their intermediate values of $(L\ S)$, respectively $(1\ 0)$ and $(1\ 1)$, the singlet state and the triplet state. In fact, in this case S is not quite "a good quantum number.". The states that are eigenstates of the Hamiltonian $\hat{H}_A$, of $\hat{\mathbf{J}}^2$ and of $\hat{J}_z$ are not eigenstates of $\hat{\mathbf{S}}^2$; the consequence of this is that the two states commonly designated as $6s6p\ ^1P_1$ and $6s6p\ ^3P_1$ are, in fact, slightly mixed.

This mixture is further discussed briefly in Section 7.7 (in Section iii on the selection rules for spin).

It is hoped that the brief discussion of the two examples, a one-electron atom and a two-electron atom, will illustrate some of the features of state labeling. We shall continue by writing simply $|\alpha J M\rangle$, with the label α containing (and concealing) all the information about the configuration and the coupling that goes to produce the final JM state. Occasionally, as necessary, we may specify the state a little more accurately by including some information on the coupling, as either $|\alpha, (L S), J M\rangle$ or $|\alpha J M, (^{2S+1}L_J)\rangle$.

7.6. THE MATRIX ELEMENTS OF AN OPERATOR

A quantum-mechanical operator $\hat{A}$, corresponding to some physical quantity of the system, can be represented by a set of quantities $\langle\alpha' J' M'|\hat{A}|\alpha J M\rangle$ with respect to a chosen complete set of unit state vectors (and their conjugates) used to describe the states of the system. These quantities are called the *matrix elements* of $\hat{A}$ in the chosen representation (in this case, αJM). For the operators already introduced the matrix elements are simple:

$$\begin{aligned}
\langle\alpha' J' M'|\hat{H}_A|\alpha J M\rangle &= W_{\alpha J}\delta(\alpha'\alpha)\delta(J'J)\delta(M'M) \\
\langle\alpha' J' M'|\hat{\mathbf{J}}^2|\alpha J M\rangle &= J(J+1)(h/2\pi)^2\delta(\alpha'\alpha)\delta(J'J)\delta(M'M) \\
\langle\alpha' J' M'|\hat{J}_z|\alpha J M\rangle &= M(h/2\pi)\delta(\alpha'\alpha)\delta(J'J)\delta(M'M) \\
\langle\alpha' J' M'|\hat{H}_{\text{mag}}|\alpha J M\rangle &= g\mu_B BM\delta(\alpha'\alpha)\delta(J'J)\delta(M'M)
\end{aligned} \tag{7.22}$$

i.e., the only nonzero elements are the diagonal ones where $\alpha' = \alpha$, $J' = J$, $M' = M$, and these matrix elements are the eigenvalues of the respective operators. We have a class of operators, which all commute with each other, representing certain physical observable quantities of the atom which are compatible, i.e., the allowed values of these observables can be simultaneously specified.

There are, however, a number of observable quantities that cannot simultaneously be specified along with the eigenvalues of energy, square of (or modulus of) angular momentum, z component of angular momentum; their quantum-mechanical operators do not commute with $\hat{H}_A$, $\hat{\mathbf{J}}^2$, $\hat{J}_z$. One such is $\hat{\mathbf{J}}$ itself. An important class of such operators is the vector operator

$\hat{\mathbf{T}}$, which has the following commutation rule with respect to $\hat{\mathbf{J}}$ ($\hat{\mathbf{J}}$ has this property with respect to itself):

$$[\hat{\mathbf{J}}, \hat{\mathbf{T}}] = \hat{\mathbf{J}}\hat{\mathbf{T}} - \hat{\mathbf{T}}\hat{\mathbf{J}} = \frac{-ih}{2\pi}\hat{\mathbf{T}} \times \mathbf{I}$$

where $\mathbf{I} = \mathbf{ii} + \mathbf{jj} + \mathbf{kk}$, the idem tensor. This is a short way of writing the commutation relationships between the components of $\hat{\mathbf{J}}$ and $\hat{\mathbf{T}}$:

$$\begin{pmatrix} [\hat{J}_x, \hat{T}_x] & [\hat{J}_x, \hat{T}_y] & [\hat{J}_x, \hat{T}_z] \\ [\hat{J}_y, \hat{T}_x] & [\hat{J}_y, \hat{T}_y] & [\hat{J}_y, \hat{T}_z] \\ [\hat{J}_z, \hat{T}_x] & [\hat{J}_z, \hat{T}_y] & [\hat{J}_z, \hat{T}_z] \end{pmatrix} = \frac{-ih}{2\pi} \begin{pmatrix} 0 & -\hat{T}_z & \hat{T}_y \\ \hat{T}_z & 0 & -\hat{T}_x \\ -\hat{T}_y & \hat{T}_x & 0 \end{pmatrix}.$$

This commutation rule applies to a large class of vectors met with in atomic physics; a list is given in Condon and Shortly,[1] p. 59. We shall mention the vector operators for the total angular momentum $\hat{\mathbf{J}}$, the position vector of any electron $\hat{\mathbf{r}}$, and the electric dipole moment of the atom $\hat{\mathbf{D}} = -e \sum_i \hat{\mathbf{r}}$, all of which are of type $\hat{\mathbf{T}}$.

The Matrix Elements of a Vector Operator of Type $\hat{\mathbf{T}}$

The vector operator $\hat{\mathbf{T}}$ can be expanded in components along a chosen set of orthogonal axes, specified by the unit vectors $\mathbf{x}^0, \mathbf{y}^0, \mathbf{z}^0$:

$$\hat{\mathbf{T}} = \mathbf{x}^0\hat{T}_x + \mathbf{y}^0\hat{T}_y + \mathbf{z}^0\hat{T}_z$$

But it is often more convenient, especially when describing states of polarization, as discussed in Section 3.2, to expand in terms of complex unit vectors:

$$\begin{aligned} \hat{\mathbf{T}} &= \frac{\mathbf{x}^0 - i\mathbf{y}^0}{\sqrt{2}} \frac{\hat{T}_x + i\hat{T}_y}{\sqrt{2}} + \mathbf{z}^0\hat{T}_z + \frac{\mathbf{x}^0 + i\mathbf{y}^0}{\sqrt{2}} \frac{\hat{T}_x - i\hat{T}_y}{\sqrt{2}} \\ &= \mathbf{e}^-\hat{T}_{+1} + \mathbf{e}^0\hat{T}_0 + \mathbf{e}^+\hat{T}_{-1} \end{aligned} \tag{7.23}$$

where $\mathbf{e}^{\pm} = (1/\sqrt{2})(\mathbf{x}^0 \pm i\mathbf{y}^0)$, $\mathbf{e}^0 = \mathbf{z}^0$ (these definitions differ in sign convention from the vectors often introduced, but they are simple and convenient to use for the present purpose). The complex components $\hat{T}_q$ ($q = -1, 0, +1$) are a special case of the components of a more general tensor operator $\hat{T}_q^k$ (k having integer values, and $q = -k, -k+1, \ldots, 0, k-1, +k$). $\hat{T}_q$, with $k = 1$, is a rank-1 tensor or vector operator.

The matrix elements of $\hat{T}_q$ in the αJM representation can be expanded as [see, for example, Edmonds,[3] Eq. (5.4.1)]:

$$\langle \alpha' J' M' | \hat{T}_q | \alpha J M \rangle = (-1)^{J'-M'} \begin{pmatrix} J & J' & 1 \\ M & -M' & q \end{pmatrix} \langle \alpha' J' \| \hat{T} \| \alpha J \rangle \tag{7.24}$$

This factorization separates out the dependence on M, M', q. The factor $\begin{pmatrix} J & J' & 1 \\ M & -M' & q \end{pmatrix}$ is a Wigner 3-j symbol which has the general form $\begin{pmatrix} j_1 & j_2 & j_3 \\ m_1 & m_2 & m_3 \end{pmatrix}$ in which $j_3 = k = 1$, $m_3 = q = -1, 0, +1$. The values of $\begin{pmatrix} J & J' & 1 \\ M & -M' & q \end{pmatrix}$ appropriate to the present discussion can be found from the following four special algebraic values:

$$\text{(i)} \quad \begin{pmatrix} j & j-1 & 1 \\ m & -m-1 & 1 \end{pmatrix} = (-1)^{j-m} \frac{[(j-m-1)(j-m)]^{1/2}}{[(2j-1)2j(2j+1)]^{1/2}}$$

$$\text{(ii)} \quad \begin{pmatrix} j & j-1 & 1 \\ m & -m & 0 \end{pmatrix} = (-1)^{j-m} \frac{[2(j-m)(j+m)]^{1/2}}{[(2j-1)2j(2j+1)]^{1/2}}$$

$$\text{(iii)} \quad \begin{pmatrix} j & j & 1 \\ m & -m-1 & 1 \end{pmatrix} = (-1)^{j-m} \frac{[2(j-m)(j+m+1)]^{1/2}}{[2j(2j+1)(2j+2)]^{1/2}} \tag{7.25}$$

$$\text{(iv)} \quad \begin{pmatrix} j & j & 1 \\ m & -m & 0 \end{pmatrix} = (-1)^{j-m} \frac{2m}{[2j(2j+1)(2j+2)]^{1/2}}$$

In order to cover all the possibilities, the following symmetry rules are required: for the 3-j symbol $\begin{pmatrix} j_1 & j_2 & j_3 \\ m_1 & m_2 & m_3 \end{pmatrix}$

i. An even permutation of the columns leaves the value unchanged;
ii. An odd permutation of the columns multiplies the value by $(-1)^{j_1+j_2+j_3}$,
iii. Changing the sign of *every* m in the bottom row multiplies the value by $(-1)^{j_1+j_2+j_3}$.

The properties of the 3-j symbol, which express the coupling of three angular momenta j_1, j_2, j_3 and their z-components m_1, m_2, m_3, are as follows:

i. The vectors specified by j_1, j_2, j_3 must form a triangle of vectors. Applied to the present case with $j_3 = k = 1$, this means that, for a given value of J, the only allowed values of J' are $J' = J - 1, J, J + 1$.
ii. In particular, J and J' cannot both be zero.
iii. The sum of the components $m_1 + m_2 + m_3 = 0$. Applied to the present case this means that $M' = M + q = M - 1, M, M + 1$.

These properties of the 3-j symbol immediately establish some of the selection rules for atomic transitions. They are discussed in the next section.

The remaining factor in the reduction of the matrix element, Eq. (7.24), is $\langle \alpha' J' \| \hat{T} \| \alpha J \rangle$, called the reduced matrix element; it is independent of the orientation in space of the vector operator $\hat{\mathbf{T}}$. The reduced matrix element still depends on the spins of the states involved through $\mathbf{J} = \mathbf{L} + \mathbf{S}$. It is possible to further factorize to separate out the S (spin) dependence. If $\hat{\mathbf{T}}$ works only on the $\mathbf{L}$ part of $\mathbf{J}$ and not on the $\mathbf{S}$ part (such would be the case for the dipole operator $\hat{\mathbf{D}}$ or the displacement operator $\hat{\mathbf{r}}$), then [see Edmonds,[3] Eq. (7.1.7)] we have

$$\langle \alpha' \, L' \, S \, J' \| \hat{T} \| \alpha \, L \, S \, J \rangle = (-1)^{L'+S+J+1}[(2J'+1)(2J+1)]^{1/2} \times \begin{Bmatrix} L' & J' & S \\ J & L & 1 \end{Bmatrix} \langle \alpha' \, L' \| \hat{T} \| \alpha \, L \rangle \tag{7.26}$$

All the dependence on J and S has been removed from the matrix element. The symbol in curly brackets is the 6-j symbol $\begin{Bmatrix} j_1 & j_2 & j_3 \\ j_4 & j_5 & j_6 \end{Bmatrix}$, whose properties and values are defined in Edmonds.[3] The remaining reduced matrix element $\langle \alpha' \, L' \| \hat{T} \| \alpha \, L \rangle$ depends only on the radial (orbital) parts of the wave functions of the two states. The value of the reduced matrix element can be deduced if the functional forms of the radial wave functions are known; these are known in analytic forms only for the hydrogenic atoms. This will be further discussed in Section 7.10 and in Appendix 6. However, even if the exact values of the matrix elements cannot be found, this reduced matrix element will form a common factor for a set of atomic transitions between states of common configurations and orbital structures, the components within the fine structure of a general transition. The *relative* strengths and lifetimes of these components can then be predicted.

In this section the Wigner 3-j and 6-j symbols have been introduced. Extensive tables of values are published in Rotenburg, Bivins, Metropolis and Wooten.[4]

7.7. SELECTION RULES FOR ELECTRIC DIPOLE RADIATIVE TRANSITIONS

We shall confine discussion here to the strongest type of radiative transition—the *electric dipole* transition. There are other types of radiative transitions which have their classical counterpart in magnetic dipole, electric quadrupole, and higher multipole charge and current distributions.

The classical electric dipole is formed by a charge—e.g., an electron $(-e)$—with position vector $\mathbf{r}$, where $\mathbf{r}$ is undergoing some cyclic oscillation. The classical electric dipole moment is $\mathbf{D} = -e\mathbf{r}$. In Chapters 2 and 3 we discussed the radiation from such a moving charge and showed, in Eq. (3.31), that the total electromagnetic power radiated from an oscillating charge is

$$P_{\text{rad}} = \frac{\mu_0}{4\pi} \frac{32\pi^4 \nu_0^4}{3c} e^2 \mathbf{A}_0 \cdot \mathbf{A}_0 = \frac{\mu_0}{4\pi} \frac{32\pi^4 \nu_0^4}{3c} e^2 \mathbf{D}_0 \cdot \mathbf{D}_0$$

where $\mathbf{A}_0$ is the rms amplitude of the displacement vector $\mathbf{r}$. The quantum-mechanical operator corresponding to the dipole moment is

$$\hat{\mathbf{D}} = -e\hat{\mathbf{r}}$$

where $\hat{\mathbf{r}}$ is the displacement operator. In the quantum-mechanical treatment, the rms amplitude of the dipole moment will be replaced with the matrix element (see Section 7.9):

$$\sqrt{2}\langle \alpha' J' M' | \hat{\mathbf{D}} | \alpha J M \rangle = -e\sqrt{2}\langle \alpha' J' M' | \hat{\mathbf{r}} | \alpha J M \rangle$$

The evaluation of this is done using the methods of the last section.

We can immediately see that, unless the following selection rules are obeyed, the matrix element is zero, the power radiated is zero, the transition is not allowed. There are three categories of selection rule:

i. Angular Momentum. In a transition, according to the quantum rule established in Chapter 6, the angular momentum of the atom must change by one unit. The general rule expressing this is

$$\Delta J = \pm 1, 0, \qquad \Delta M = \pm 1, 0 \tag{7.27}$$

The relative probabilities of the transitions are determined through the values of the 3-j symbols. Since the angular momentum *must* change by one unit, the following conditions apply: if $J' = J$, the M' and M must differ by unity, $\Delta M = \pm 1$; and $\Delta M = 0$ is not allowed. This further means that a transition from $J' = 0 \leftrightarrow J = 0$ is not allowed. These conditions are contained in the values of the 3-j symbols.

ii. Parity. The operator $\hat{\mathbf{r}}$ (and the operator $\hat{\mathbf{D}}$) has odd parity; i.e., it reverses sign on reflection of the spatial coordinates. Consequentially, since the evaluation of the matrix element of the dipole transition involves a spatial integration over *all* space (see Section 7.10), the value of the matrix element will be zero unless the integrand is an even function of spatial coordinates. Therefore the two states involved must be of different parity. The parity of an atomic state is determined by the sum $l_1 + l_2 + l_3 + \cdots$ of all the electrons in the configuration; if the sum is odd, the parity of the state is odd; if the sum is even, the parity of the state is even. For the case of a transition where only one electron is involved, the parity changes if, for that electron,

$$\Delta l = \pm 1 \tag{7.28}$$

This is the requirement on the change in the configuration in transition. In the simplest cases, this leads to the rule that $\Delta L = \pm 1$ for the whole atom; however, in complicated atoms, this is not always true. The essential requirement is that one of the states must have a configuration of odd parity, and the other must have even parity.

iii. Spin. An electric dipole transition involves only the change of orbit; the spin of the electron does not change. Therefore we have

$$\Delta S = 0 \tag{7.29}$$

This would lead to the situation, for example, that in two-electron atoms that have singlet states ($S = 0$) and triplet states ($S = 1$), transitions are only allowed between those states that have the same multiplicity. One should note, however, that S is not always a good quantum number; certain allowed states of the atom, being eigenstates of the Hamiltonian (and simultaneously of $\hat{\mathbf{J}}^2$ and $\hat{J}_z$), are mixtures of two states of different S quantum number. A well-known example occurs in mercury. As discussed in Section 7.5 there are two different states with $J = 1$ in the $6s6p$ configuration, the first excited states of mercury. One is the so-called "singlet" state $|6s6p\ (L=1\ S=0)\ J=1\ M\rangle$ or $6s6p$ "$^1P_1$" state; the other is the so-called "triplet" state $|6s6p\ (L=1\ S=1)\ J=1\ M\rangle$ or $6s6p$ "$^3P_1$" state. The latter cannot strictly make a transition to the ground state, $|6s^2\ (L=0\ S=0)\ J=0\ M=0\rangle$ or $6s^2\,{}^1S_0$, because $\Delta S = 1$ is not allowed. However, this transition is observed to occur, being the well-known and quite strong 253.7-nm line in the ultraviolet. The reason is that this level, commonly referred to as a triplet state "$^3P_1$" is really a mixture—largely the pure $6s6p\,{}^3P_1$ but with a small admixture of $6s6p\,{}^1P_1$. This is why the state symbol has been written above in inverted commas. Similarly, the level commonly referred to as a singlet state "$^1P_1$" (with a 185.0-nm line in transition to ground) is largely the pure $6s6p\,{}^1P_1$ but with a small admixture of $6s6p\,{}^3P_1$. Such admixtures can occur when there are several states that derive from the same configuration and have the same value of total angular momentum specified by J. We note that the other two states of this configuration, $6s6p\,{}^3P_0$ with $J = 0$ and $6s6p\,{}^3P_2$ with $J = 2$, are *pure* triplet states. They cannot decay to the ground for several reasons. Firstly, being pure triplet, they cannot decay to the pure singlet ground. The $^3P_0$ is also forbidden because it would be $J' = 0$ to $J = 0$, which is strictly forbidden. The $6s6p\,{}^3P_2$ would be weak since $\Delta J = 2$, an electric quadrupole transition; but $\Delta S = 1$ makes it completely forbidden. These two states cannot decay to ground; they are truly *metastable.* Atoms that reach these states by cascade transitions from upper levels cannot decay by radiation. The atoms remain in these states until they are released by collisions with the walls or with other atoms.

The fact is that $\Delta S = 0$ is a perfectly good selection rule, but states that are frequently labeled with an S quantum number are sometimes not pure in that label.

Equations (7.27), (7.28), and (7.29) state the electric dipole section rules for atomic transitions.

7.8. SUPERPOSITION STATES AND THE EQUATION OF MOTION

So far we have dealt with *stationary* states, eigenstates of the atomic Hamiltonian. These states we shall label simply as $|0\rangle, |1\rangle, |2\rangle, \ldots, |n\rangle, \ldots$ in order of their energies (we introduce now the label 0 for the ground state, which we avoided earlier in this chapter). The lowest state has energy W_0, which we may, without loss of generality, regard as zero. The excited states $|1\rangle, |2\rangle, \ldots$ have energies $W_1 = h\nu_1$, $W_2 = h\nu_2, \ldots$ etc. They are eigenstates of the atomic Hamiltonian $\hat{H}_A$, such that $\hat{H}_A|j\rangle = W_j|j\rangle$. All states are more fully labeled with their configuration α, with the angular momentum, quantum numbers J and M, and any other information needed to specify them.

However, an atom may be in a more general state than one of these eigenstates; it may be changing its state with time, either by decaying naturally from one state to another, or being stimulated by outside perturbations to make transitions. Thus the general state may be some superposition of the eigenstates, and there will be some equation of motion that will predict how the general state changes with time.

A general equation of motion for a quantum system, called the time-dependent Schrödinger equation, and which we shall simply state, is

$$\frac{ih}{2\pi}\frac{\partial}{\partial t}|\ \rangle = \hat{H}|\ \rangle \tag{7.30}$$

where $\hat{H}$ is a general Hamiltonian for the system involving all the internal interactions and those with its surroundings that lead to the total energy. The ket represents a general state of the system, a superposition of eigenstates with, in general, time-dependent coefficients. Let us first apply the equation to "stationary" states with the atomic Hamiltonian $\hat{H}_A$. These states obviously have some time-dependent factor associated with them, so we shall write them as $|n, t\rangle$, the time-dependent form of the stationary state $|n\rangle$:

$$\frac{ih}{2\pi}\frac{\partial}{\partial t}|n, t\rangle = \hat{H}_A|n, t\rangle = W_n|n, t\rangle$$

$$\frac{\partial}{\partial t}|n, t\rangle = \frac{-i2\pi W_n}{h}|n, t\rangle \tag{7.31}$$

$$|n, t\rangle = e^{-i2\pi W_n t/h}|n, 0\rangle = e^{-i2\pi\nu_n t}|n\rangle$$

That is, associated with each *stationary* eigenstate of the Hamiltonian $\hat{H}_A$ is a phase factor $e^{-i2\pi W_n t/h} = e^{-i2\pi\nu_n t}$, which defines the energy of the state. The *time-dependent* form of the *stationary* state can be written $|n, t\rangle = e^{-i2\pi\nu_n t}|n\rangle$.

The general superposition state in which an atom may be found is some sort of superposition of the possible eigenstates of the system:

$$|t\rangle = \cdots a_i e^{-i2\pi\nu_i t}|i\rangle + a_j e^{-i2\pi\nu_j t}|j\rangle + a_k e^{-i2\pi\nu_k t}|k\rangle \cdots \tag{7.32}$$

The coefficients ... a_i, a_j, a_k, ... are called the *probability amplitudes* of the eigenstates. The general state is written with a t inside the ket because, quite apart from the time-dependent phases that describe the energies of the eigenstates, the amplitudes ascribed to each state may themselves depend on time. In other words, the general state described by $|t\rangle$ is evolving in time.

Let us assume that any state described by a ket vector is normalized—a sort of conservation law saying that the atom cannot disappear, only change its state. Therefore, using the orthonormality condition for eigenstates, Eq. (7.8), we obtain

$$\langle t|t\rangle = 1 = \cdots a_i^* a_i + a_j^* a_j + a_k^* a_k + \cdots \tag{7.33}$$

The terms $a_n^* a_n$ obviously represent the *probability* that an atom described by the superposition Eq. (7.32) is to be found in eigenstate $|n\rangle$.

Suppose we look at the equation of motion for an atom in a general state, as expressed in Eq. (7.32), under circumstances where the energy is determined only by the internal dynamics and interactions expressed by $\hat{H}_A$. Substitute Eq. (7.32) into the equation of motion:

$$\frac{ih}{2\pi}\frac{\partial}{\partial t}|t\rangle = \hat{H}_A|t\rangle$$

Expand $|t\rangle$, carry out the differentiations and operate with $\hat{H}_A$:

$$\frac{ih}{2\pi}\dot{a}_i e^{-i2\pi\nu_i t}|i\rangle + a_i h\nu_i e^{-i2\pi\nu_i t}|i\rangle + \text{similar terms in } j, k\cdots$$

$$= a_i h\nu_i e^{-i2\pi\nu_i t}|i\rangle + \text{similar terms in } j, k\cdots \tag{7.34}$$

i.e.,

$$\frac{ih}{2\pi}\dot{a}_i\, e^{-i2\pi\nu_i t}|i\rangle + \text{similar terms in } j, k\cdots = 0$$

Multiplying successively through by $\langle i|$, $\langle j|$, etc., and, using the orthonormal rule, we obtain

$$\dot{a}_i = 0, \qquad \text{i.e., } a_i \text{ is constant in time,}$$

$$\dot{a}_j = 0, \qquad \text{i.e., } a_j \text{ is constant in time, etc.}$$

All the probability amplitudes are constant in time, the atom rests in its initial superposition of eigenstates, and the general superposition state is a stationary state of the atomic Hamiltonian. For an atom, this obviously does not correspond to reality, since excited states always decay spontaneously. Therefore, in reality, $\hat{H}_A$ can never be the only contribution to the total Hamiltonian of the atom.

Suppose, however, that the atom is subjected to some perturbing influence described by a Hamiltonian, $\hat{H}_P$, representing the energy of perturbation by interaction between the atom and its environment. (Or perhaps an extra energy of internal interaction that has not yet been allowed for. An example is the magnetic interaction between the magnetic moment of the nucleus and the internal magnetic field caused by the angular momentum **J**, which gives *hyperfine structure* on the states and the spectral lines.) Then $\hat{H} = \hat{H}_A + \hat{H}_P$ and the equation of motion becomes

$$\frac{ih}{2\pi}\dot{a}_i\, e^{-i2\pi\nu_i t}|i\rangle + \cdots + \frac{ih}{2\pi}\dot{a}_k\, e^{-i2\pi\nu_k t}|k\rangle\cdots$$

$$= a_i\, e^{-i2\pi\nu_i t}\hat{H}_P|i\rangle + \cdots + a_k\, e^{-i2\pi\nu_k t}\hat{H}_P|k\rangle\cdots \tag{7.35}$$

where cancellation of certain terms has been carried out, and the terms in i and k only have been retained in order to illustrate the behavior. In order to isolate the differential $\dot{a}_i$, multiply through by $(-i2\pi/h)\, e^{+i2\pi\nu_i t}\langle i|$:

$$\dot{a}_i = \frac{-i2\pi}{h}\langle i|\hat{H}_P|i\rangle a_i + \frac{-i2\pi}{h}\langle i|\hat{H}_P|k\rangle\, e^{-i2\pi(\nu_k-\nu_i)t}a_k\cdots \tag{7.36}$$

and similar differential equations for the other amplitudes. We can now consider two circumstances.

i. The Perturbing Hamiltonian Commutes with the Atomic Hamiltonian. In this case $\hat{H}_P$ has only diagonal elements, in which case Eq. (7.36) reduces to

$$\dot{a}_i = \frac{-i2\pi}{h} \langle i|\hat{H}_P|i\rangle a_i \tag{7.37}$$

and similar equations for the other amplitudes. If $\hat{H}_P$ is independent of time, this has the solution

$$a_i(t) = a_i(0) \exp\left(\frac{-i2\pi}{h} \langle i|\hat{H}_P|i\rangle t\right) = a_i(0)\, e^{-i2\pi f_{ip} t} \tag{7.38}$$

where $f_{ip} = \langle i|\hat{H}_P|i\rangle / h$. The general state can be written as

$$|t\rangle = \cdots a_i\, e^{-i2\pi(\nu_i + f_{ip})t}|i\rangle$$
$$+ a_j\, e^{-i2\pi(\nu_j + f_{jp})t}|j\rangle + a_k\, e^{-i2\pi(\nu_k + f_{kp})t}|k\rangle \cdots \tag{7.39}$$

The energy $h\nu_i$ associated with each state has been shifted to

$$h(\nu_i + f_{ip}) = h\nu_i + \langle i|\hat{H}_P|i\rangle$$

An example of this type of perturbation is $\hat{H}_P = -\hat{\mathbf{m}} \cdot \mathbf{B}$, representing the energy of interaction between the magnetic dipole moment of the atom $\mathbf{m} = -g\mu_B \mathbf{J}/h$ and an applied magnetic field $\mathbf{B}$. The perturbing Hamiltonian becomes $\hat{H}_P = g\mu_B \hat{\mathbf{J}} \cdot \mathbf{B}/h = g\mu_B \hat{J}_z B/h$, and the matrix element is

$$\langle \alpha_i J_i M_i |\, \hat{H}_P \,| \alpha_i J_i M_i\rangle = g\mu_B B M_i \tag{7.40}$$

The energy level represented by $h\nu_i$ is split into $2J_i + 1$ components with energies $h\nu_i + g\mu_B B M_i$. The degeneracy of the levels with various values of M_i has been removed by the presence of the magnetic field $\mathbf{B}$ which makes the space nonisotropic. The splitting of the state of the atom into several substates of different energies leads to a splitting of the spectral lines in which this state is involved into several components. This is the Zeeman effect.

ii. The Perturbing and Atomic Hamiltonians Do Not Commute. Suppose that $\hat{H}_P$ does not commute with $\hat{H}_A$ but has properties such that some other matrix element $\langle i|\hat{H}_P|k\rangle$ is nonzero—such as would happen if $\hat{H}_P$ were an operator of type $\hat{\mathbf{T}}$ (see Section 7.6). In this case Eq. (7.36) reduces to

$$\dot{a}_i = \frac{-i2\pi}{h} \langle i|\hat{H}_P|k\rangle\, e^{-i2\pi(\nu_k - \nu_i)t} a_k \cdots \tag{7.41}$$

and we have also the coupled equation:

$$\dot{a}_k = \frac{-i2\pi}{h} \langle k|\hat{H}_P|i\rangle\, e^{+i2\pi(\nu_k - \nu_i)t} a_i \cdots \tag{7.42}$$

In general there may be more than one nonzero term on the right-hand sides. Such would be the case for the static Stark effect where the perturbation $\hat{H}_P = -\hat{\mathbf{D}} \cdot \mathbf{E}$ represents the interaction of the electric dipole moment of the atom with a static applied electric field $\mathbf{E}$. If we were interested in the effect on the state $|i\rangle$, we would have to include all states $|k\rangle$ on the right-hand side of Eq. (7.41) and for the coupled equations (7.42) that have nonzero matrix elements $\langle i|\hat{H}_P|k\rangle$. We shall not study this case further. However, the simpler pair of coupled equations (7.41) and (7.42) are important in a case we shall treat in later chapters. This is where the perturbation is time dependent, containing frequencies close to *one* of the state difference frequencies $\nu_k - \nu_i$. The effect of resonance restricts the right-hand side of Eqs. (7.41) and (7.42) to one term each. This is the time-dependent perturbation case that leads to transitions between states, sometimes also referred to as the dynamic Stark effect. We shall leave the study of this important case to later chapters.

A further case of importance is that of a perturbation due to an interaction expressed by $\hat{H}_P$, which leads to the decay of excited states; this case is treated in Chapter 10.

7.9. THE CORRESPONDENCE PRINCIPLE

In the early days of the development of quantum mechanics it was found fruitful to pass from classical variables to a quantum formulation by the use of a *correspondence principle.* The matter is discussed in Condon and Shortley.[1] For the purposes required here the rms amplitude of an oscillating quantity (such as the dipole moment amplitude $\mathbf{D}_0 = -e\mathbf{A}_0$) is replaced by the matrix element of the appropriate quantum-mechanical operator according to

$$\mathbf{D}_0 \Rightarrow \sqrt{2}\langle i|\hat{\mathbf{D}}|k\rangle \tag{7.43}$$

$\mathbf{D}_0$ is the rms amplitude of the equivalent classical oscillator responsible for the particular frequency in the atom; $|i\rangle$ and $|k\rangle$ are the state vectors involved in the quantum transition responsible for the radiation at that frequency; $\hat{\mathbf{D}} = -e\Sigma\hat{\mathbf{r}}$ is the quantum-mechanical operator.

7.10. THE WAVE FUNCTION

In the treatment so far given we have described the state of the atom in terms of orthonormal state vectors $|\alpha J M\rangle$ which are simultaneous eigenvectors of the atomic Hamiltonian $\hat{H}_A$ and other appropriate operators that commute with $\hat{H}_A$, such as $\hat{\mathbf{J}}^2, \hat{J}_z, \ldots$. This description labels the state with only those quantities whose values are known for the state, its configuration, its total angular momentum, one component of the angular momentum, etc. In principle, the energy is also specified because the state is an eigenstate of the Hamiltonian, but its value cannot, in general, be deduced from the theory.

There is an alternative, and entirely equivalent, way of writing the state of the atom: by use of the *wave function* ψ for the state. This was the method used in the early days of quantum mechanics and it is still widely used. The general algebra of the quantum-mechanical wave function is discussed in books on quantum mechanics (e.g., Cohen-Tannoudji, Diu, and Laloë[2]). We shall here quote only a few appropriate properties. The wave function is a function of the coordinates of space and perhaps of time: $\psi = \psi(x, y, z; t) = \psi(r, \theta, \phi; t)$. It has the following significance for the electron state: $\psi^*\psi$ is the probability that the electron will be found per unit volume at that point in space at that time. Since the electron must be somewhere in space we have the normalization

$$\int_V \psi^*\psi \, dx \, dy \, dz = \int_V \psi^*\psi r^2 \sin\theta \, dr \, d\theta \, d\phi = 1 \tag{7.44}$$

where the volume integration is over all space. If ψ is the wave function representing an eigenstate of the atomic Hamiltonian, and of other operators of observables, it can be labeled with the appropriate quantum numbers—$\alpha, L, M_L, S, M_S, J, M$—required to define the state. In general the wave function will be a function of spin-space as well as of coordinate space. However, for many purposes, it is appropriate to think of the wave function in coordinate space only; then it will depend only on the configuration and the orbital specifications of the atomic state.

It is seen that the wave function implies more than the state-vector—it implies that the spatial distribution of the state of the system is known. However, it is only in certain circumstances, such as for the hydrogenic atom, that this is known for atomic states.

For some physical quantity Q, scalar, vector or tensor, which has a quantum operator equivalent $\hat{Q}$, there is a matrix element

$$\langle \alpha' J' M' | \hat{Q} | \alpha J M \rangle = \int \psi^*_{\alpha' J' M'} \hat{Q} \psi_{\alpha J M} \, d\tau \tag{7.45}$$

If $\alpha', J', M' = \alpha, J, M$ then the quantity

$$\langle \alpha\, J M | \hat{Q} | \alpha\, J M \rangle = \int \psi^*_{\alpha JM} \hat{Q} \psi_{\alpha JM}\, d\tau \tag{7.46}$$

is called the *expectation value* of the physical quantity Q when the system is in the state $|\alpha\, JM\rangle$. As can be seen from the right-hand side of Eq. (7.46) it is the mean value taken over the distribution of the state expressed by the wave function ψ. In the expressions $d\tau$ represents an element of coordinate and spin space.

We shall consider two particular examples of wave functions.

i. The Wave Function of a Free Particle

A free particle of linear momentum $\mathbf{p} = p\mathbf{k}^0$ ($\mathbf{k}^0$ is the unit vector defining the propagation direction) has a wave function

$$\psi = e^{+i2\pi p\mathbf{k}^0\cdot\mathbf{r}/h} = e^{+i2\pi\mathbf{k}^0\cdot\mathbf{r}/\lambda} \tag{7.47}$$

where $\lambda = h/p$ is the de Broglie wavelength. When this is combined with the temporal factor $e^{-i2\pi\nu t}$, where $h\nu$ is the energy of the free particle, we obtain

$$e^{-i2\pi\nu t}\psi = e^{-i2\pi(\nu t - \mathbf{k}^0\cdot\mathbf{r}/\lambda)} \tag{7.48}$$

This represents a plane wave of frequency ν and of wavelength λ traveling in the $\mathbf{k}^0$ direction. The free particle has a total energy (kinetic + rest mass) of $h\nu = mc^2$ and a linear momentum of $p = h/\lambda = mv$, where, in these expressions, $m = \gamma m_0$ is the relativistic mass. The phase velocity of the wave is $u = \nu\lambda = c^2/v$. The velocity of the particle = v = the group velocity of the wave.

An alternative way of establishing the wave function of the free particle is the following (using nonrelativistic analysis). The kinetic energy of a particle is connected to its linear momentum by the relationship $K = p^2/2m$. Written in terms of quantum operators this becomes

$$\hat{H}_{\text{free}} = \frac{\hat{p}^2}{2m} = -\frac{h^2}{8\pi^2 m}\nabla^2$$

Here we have introduced the quantum-mechanical operator for the momentum $\hat{\mathbf{p}} = -(ih/2\pi)\boldsymbol{\nabla}$. This is the equivalent of the relation for the Hamiltonian $\hat{H} = (ih/2\pi)(\partial/\partial t)$ introduced in Eq. (7.30). The equation $\hat{H}_{\text{free}}\psi = K\psi$ then becomes, for a free particle of kinetic energy $K = \frac{1}{2}mv^2$,

$$\nabla^2\psi = -\frac{8\pi^2 m}{h^2}\,\frac{1}{2}mv^2\psi = -\left(\frac{2\pi mv}{h}\right)^2\psi = -\left(\frac{2\pi}{\lambda}\right)^2\psi$$

with the solution $\psi = e^{+i2\pi\mathbf{k}^0\cdot\mathbf{r}/\lambda}$. This wave function is appropriate to a particle per unit volume; if the particle exists in a box (taken to be cubical of dimension L), the wave function would be

$$\psi = \frac{1}{L^3} e^{+i2\pi\mathbf{k}^0\cdot\mathbf{r}/\lambda} = \frac{1}{L^3} e^{+i2\pi\sigma\mathbf{k}^0\cdot\mathbf{r}} \tag{7.49}$$

ii. The Wave Function of Atomic States

For an atom, where the electron is bound to the center (nucleus plus core electrons) by a potential energy $V(\mathbf{r})$, the Hamiltonian becomes $\hat{H}_A = \hat{p}^2/2m + V(\hat{\mathbf{r}})$, the quantum operator representing the sum of the kinetic and the potential energies of the electron. When this is placed in the equation of motion $\hat{H}_A\psi = W\psi$, one gets

$$\left[\frac{\hat{p}^2}{2m} + V(\hat{\mathbf{r}})\right]\psi = W\psi \tag{7.50}$$

This is called Schrödinger's time-independent equation. It is solvable only in certain cases where the potential energy operator function $V(\hat{\mathbf{r}})$ is known. The only atomic case where a complete solution can be found is for the hydrogenic atom where $V(\mathbf{r}) = -(1/4\pi\varepsilon_0)(Ze^2/r)$. The appropriate wave function to describe the atom is just the wave function of the single electron, and can be labeled by the single electron quantum numbers n, l, $m(=m_l)$. In this case it can be shown that the wave function can be separated into two factors (we do not consider a third factor which accounts for the two possible spin states):

$$\psi_{nlm} = R_{nl}(r)Y_l^m(\theta, \phi) \tag{7.51}$$

$Y_l^m(\theta, \phi)$ are the spherical harmonics common to many problems with central symmetry. $R_{nl}(r)$ are the radial wave functions, which can only be expressed in analytic forms for certain simple cases such as for the hydrogenic atoms. These functions, and some relationships between the wave function and the bra-ket notations, are discussed in Appendix 6.

REFERENCES

1. E. U. CONDON AND G. H. SHORTLEY, *The Theory of Atomic Spectra* (Cambridge University Press, Cambridge, 1953).

2. C. COHEN-TANNOUDJI, B. DIU, AND F. LALOË, *Quantum Mechanics* (John Wiley & Sons, New York, 1977).
3. A. R. EDMONDS, *Angular Momentum in Quantum Mechanics* (Princeton University Press, Princeton, New Jersey, 1960).
4. M. ROTENBURG, R. BIVINS, N. METROPOLIS, AND J. K. WOOTEN JR., *The 3-j and 6-j Symbols* (The Technology Press, Massachusetts Institute of Technology, Cambridge, Massachusetts, 1959).

CHAPTER 8

THE EINSTEIN *A* AND *B* COEFFICIENTS

8.1. POPULATIONS AND TRANSITION RATES

Consider an ensemble of identical atoms whose states are quantized with energies W_i, W_j, $W_k, \ldots$, and with statistical weights w_i, w_j, $w_k, \ldots$, respectively. The statistical weight of a level refers to the fact that it may consist of a number w of degenerate states each of which plays its part in any statistical process involving that level. For example, for an atomic level of angular momentum specified by J, there are $2J + 1$ states which are degenerate in zero magnetic field. The statistical weight of this state is $w = 2J + 1$. In general there can be other contributions to the statistical weight such as unresolved hyperfine structures. Consider an atom of two levels only, a lower level of energy W_i and an upper level of energy W_k, between which radiative transition is allowed. Let it experience an environment of radiation containing the frequency $\nu_{ki} = (W_k - W_i)/h$. According to the quantum rules of atom–radiation interaction established in Chapter 7 absorptive transitions are stimulated from level i to level k, and emissive transitions are stimulated from level k to level i. But even if no radiation is present, it is known from observation that spontaneous transitions from level k to level i also occurs.

In a situation where the system of atom and radiation exist in equilibrium, the principle of detail balancing requires that the rate of transfer from $i \rightarrow k$ must equal that from $k \rightarrow i$. There may be circumstances, such as in a laser, where pumping through a third level affects the relative populations of the states and, in many cases, collisions may affect the populations. However, the transition probabilities through radiative process are independent of the relative population of the states; therefore we can study the equilibrium under the conditions where radiation alone is responsible. Indeed, since the spectral distribution itself is not important we can presume that the equilibrium is established inside a cavity of temperature T filled with black-body radiation.

At thermal equilibrium in an environment at temperature T, the ratio of the number of atoms in the two levels is, according to the Boltzmann distribution deduced in statistical mechanics:

$$\frac{N_k}{N_i} = \frac{w_k}{w_i} e^{-(W_k - W_i)/kT} = \frac{w_k}{w_i} e^{-h\nu_{ki}/kT} \tag{8.1}$$

Figure 9 illustrates the rates for the various transition processes between the states. Einstein introduced two coefficients that describe, in a phenomenological way, the rates at which transitions take place between two atomic levels. The Einstein A coefficient measures the probability per unit time that an atom makes a transition from one state to another by *spontaneous* transition. Thus the rate of spontaneous transition from $k \to i$ is $N_k A_{ki}$. The spontaneous rate in the reverse direction is zero. The Einstein B coefficients measure the probability per unit time of *stimulating* a transition; the stimulated rates depend also on the spectral energy density, $\rho(\nu_{ki})$, i.e., the energy per unit volume per unit frequency band-width at the frequency of the transition, as defined in Eq. (5.11). Thus the rates of stimulated transition are

$$N_i B_{ik} \rho(\nu_{ki}) \qquad \text{upward from } i \to k$$

and (8.2)

$$N_k B_{ki} \rho(\nu_{ki}) \qquad \text{downward from } k \to i$$

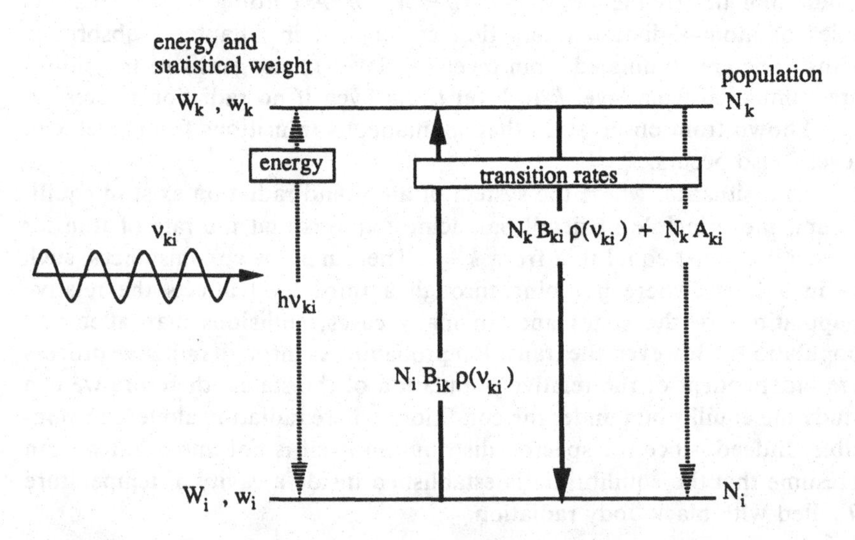

FIGURE 9. Stimulated and spontaneous transitions between two states, i and k.

Equating rates upward and downward, the ratio of the populations is obtained:

$$\frac{N_k}{N_i} = \frac{B_{ik}\,\rho(\nu_{ki})}{B_{ki}\,\rho(\nu_{ki}) + A_{ki}} \tag{8.3}$$

The Planck distribution formula for black-body radiation is:

$$\rho(\nu) = \frac{8\pi h\nu^3}{c^3}\,\frac{1}{e^{h\nu/kT} - 1} \tag{8.4}$$

[This formula is often quoted in terms of $\omega = 2\pi\nu$. Note that $\rho(\nu)\,d\nu = \rho(\omega)\,d\omega$, where $d\nu$ and $d\omega$ correspond to the same range of frequency; therefore $\rho(\nu) = 2\pi\rho(\omega)$.]

Equating the ratios of state populations as expressed in Eq. (8.1) and Eq. (8.3), using Eq. (8.4) and recognizing that the equations must be compatible at all temperatures, we obtain the following relationships between the two Einstein coefficients:

$$w_i B_{ik} = w_k B_{ki} \tag{8.5}$$

$$A_{ki} = \frac{8\pi h\nu_{ki}^3}{c^3}\,B_{ki} \tag{8.6}$$

8.2. THE CLASSICAL THEORY FOR THE A COEFFICIENT

According to Eq. (3.31) a charge $\pm e$, oscillating at frequency ν_0 and with rms amplitude A_0, radiates energy at the rate

$$P_{\text{rad}} = \frac{\mu_0}{4\pi}\,\frac{32\pi^4\nu_0^4 e^2 A_0^2}{3c} \tag{8.7}$$

We take the point of view that this applies to the radiation from an atom where $D_0 = -eA_0$ is the equivalent electric dipole moment rms amplitude of the atom. Since each atomic transition is associated with a quantum of energy $h\nu_0$ being transferred from the atom to the field (according to the quantum rules established in Chapter 6), the number of transitions per unit time, which is just the Einstein A coefficient for the transition, is

$$A_{ki} = \frac{P_{\text{rad}}}{h\nu_0} = \frac{\mu_0}{4\pi}\,\frac{32\pi^4\nu_0^3}{3hc}\,D_0^2 \tag{8.8}$$

The quantity D_0 is still, however, a classical quantity and it must be replaced by a quantity having meaning for a quantum description of the atom. We use here the correspondence principle; it has been shown in Chapter 7, Eq. (7.43), that the classical$\Rightarrow$quantum correspondence is given by $\mathbf{D}_0 \Rightarrow \sqrt{2}\langle i|\hat{\mathbf{D}}|k\rangle$, where $\hat{\mathbf{D}} = -e\hat{\mathbf{r}}$ is the dipole moment operator and $\hat{\mathbf{r}}$ is the operator corresponding to the displacement of electrons in the atom from some center in the atom. In general $\hat{\mathbf{r}} = \sum_n \hat{\mathbf{r}}_n$, the sum being taken over the electrons in the atom responsible for the transition. For the simplest first-order transitions this involves one electron only. By using this correspondence principle approach, an expression for the A coefficient is obtained:

$$A_{ki} = \frac{\mu_0}{4\pi}\frac{64\pi^4 e^2 \nu_0^3}{3ch}\langle k|\hat{\mathbf{r}}|i\rangle \cdot \langle i|\hat{\mathbf{r}}|k\rangle \tag{8.9}$$

The formula gives the probability of decay per unit time between an upper nondegenerate level k and a lower nondegenerate level i. Thus k and i are labels which should actually include the magnetic quantum numbers as well as other labels. The probability rate between nondegenerate states $|k\rangle = |J_k M_k\rangle$ and $|i\rangle = |J_i M_i\rangle$ is given by Eq. (8.9). To get the total probability rate from *any* state of $|k\rangle$ to *every* state of $|i\rangle$ we must sum Eq. (8.9) over all values of M_i. Every M_k must decay at the same rate; otherwise the spatial distribution of intensity and the polarization of the radiation would be observed to change with time. Therefore the value of A_{ki} is independent of M_k:

$$A_{ki} = \frac{\mu_0}{4\pi}\frac{64\pi^4 e^2 \nu_0^3}{3ch}\sum_{M_i}\langle J_k M_k|\hat{\mathbf{r}}|J_i M_i\rangle \cdot \langle J_i M_i|\hat{\mathbf{r}}|J_k M_k\rangle \tag{8.10}$$

This result is sometimes symmetrized by summing over all values of M_k as well as M_i and including a weighting factor $1/w_k$.

8.3. THE RADIATIVE LIFETIME AND DECAY CONSTANT

An ensemble of atoms containing $N_k(t)$ atoms in upper state at time t decays with time according to

$$\frac{dN_k(t)}{dt} = -N_k(t)\sum_i{}' A_{ki} \tag{8.11}$$

where the prime indicates that *all* levels *below* k are included in the summation. The population decays exponentially according to

$$N_k(t) = N_k(0)\, e^{-\gamma_k t} = N_k(0)\, e^{-t/\tau_k}$$

The radiative decay constant γ_k, the lifetime τ_k, and the Einstein A coefficient are related by

$$\gamma_k = \frac{1}{\tau_k} = \sum_{i<k} A_{ki} \tag{8.12}$$

It is useful to compare this quantum mechanical expression for the radiative decay constant of a level k with the decay constant of a classical oscillation given by Eq. (3.27). The classical value depends only on the frequency and if applied, for example, to the $2p \rightarrow 1s$ transition in hydrogen (Lyman-α, $\lambda = 121.6$ nm, $\nu_0 = 2.465 \times 10^{15}$ Hz), yields $\gamma_{\text{class}} = 1.503 \times 10^9\ \text{s}^{-1}$. The quantum value obtained from Eq. (8.10) gives a different result. The matrix element $\langle i|\hat{\mathbf{r}}|k\rangle = \langle J_i M_i|\hat{\mathbf{r}}|J_k M_k\rangle$ can be evaluated using the hydrogenic wave functions. Using $J_k = 1$, $M_k = 0$, $J_i = 0$, $M_i = 0$, for a typical matrix element (the others, with $M_k = \pm 1$, yield the same value), appropriate to a electron without spin:

$$\langle 0, 0|\hat{\mathbf{r}}|1, 0\rangle = \int_{r=0}^{\infty} \int_{\theta=0}^{\pi} \int_{\phi=0}^{2\pi} \psi_{2p} z \psi_{1s} r^2 \sin\theta \, dr \, d\theta \, d\phi \tag{8.13}$$

using the expressions for the hydrogenic wave functions given by Eq. (7.51) and Appendix 6 we obtain

$$\begin{aligned}
&\langle 0, 0|\hat{\mathbf{r}}|1, 0\rangle \\
&= \iiint \frac{1}{4(2\pi)^{1/2}} \frac{1}{a_B^{3/2}} \frac{r}{a_B} e^{-r/2a_B} r^3 \frac{1}{\sqrt{\pi}} \frac{1}{a_B^{3/2}} e^{-r/a_B} \cos^2\theta \sin\theta \, dr \, d\theta \, d\phi \\
&= \frac{2^8 a_B}{3^5\sqrt{2}} = 0.745 a_B
\end{aligned} \tag{8.14}$$

The Bohr radius is

$$a_B = \left(\frac{\mu_0}{4\pi}\right)^{-1} \frac{h^2}{4\pi^2 m e^2 c^2} = 5.292 \times 10^{-11}\ \text{m}$$

Substituting this value for the matrix elements in Eq. (8.12) one calculates

$$\gamma_k = A_{ki} = 6.249 \times 10^8\ \text{s}^{-1}$$

This yields a mean lifetime of $\tau_k = 1.600 \times 10^{-9}$ s, in good agreement with the measured value of $(1.600 \pm 0.004) \times 10^{-9}$ s.

Spectroscopists have traditionally compared the decay rate for a transition with the classical decay constant, writing

$$\gamma = A_{ki} = \left(\frac{3w_i}{w_k} f_{ik}\right) \gamma_{\text{class}} \tag{8.15}$$

The dimensionless quantity in parentheses is so defined because the so-called "classical" transition ($J_k = 1 \to J_i = 0$, which exhibits the normal Zeeman effect, and for which $w_k = 3$ and $w_i = 1$, i.e., for which $3w_i/w_k = 1$) often has a measured γ approaching the classical value. One expects then a value of f_{ik} close to, but less than, unity. Applying it to the Lyman-α transition just discussed, $f_{ik} = 0.416$. The quantity f_{ik} is called the "f value" of the transition.

The order of subscripts on f imply an absorption process. For the emission process

$$w_k f_{ki} = -w_i f_{ik}$$

Thus the f value for emission is negative. Note that Eq. (8.15) can be written

$$A_{ki} = (-3f_{ki})\gamma_{\text{class}} \tag{8.16}$$

There are some sum rules on f:

i. If $|i\rangle$ is a ground state:

$$\sum_k f_{ik} = 1 \tag{8.17}$$

for a spectrum that can be attributed to a single electron.

ii. If $|j\rangle$ is a general level, with levels k above it, and levels i below it:

$$\sum_k f_{jk} + \sum_i f_{ji} = 1 \tag{8.18}$$

where the f_{ji} of the second sum are negative. For completeness, the sum over the higher levels must include contributions from the continuum above the ionization level.

iii. For spectra involving more than one electron the right-hand side in these sum rules must be replaced by z, the number of equivalent electrons.

The theory leading to Eq. (8.10) for A_{ki} can hardly be described as "quantum mechanical" even though the final result is expressed in quantum-mechanical terms. It is not even semiclassical, which would start from the quantum-mechanical equation of motion Eq. (7.30), with the Hamiltonian $\hat{H}$ being the sum of $\hat{H}_A$, describing the energy intrinsic to the *atomic structure*, and a perturbation $\hat{H}_{\text{int}}$ describing the *interaction* with an electromagnetic field. Such a theory would lead to simulated absorption and emission as well as, hopefully, to the possibility of spontaneous decay. The semiclassical perturbation theory is to be discussed in subsequent chapters.

8.4. THE CLASSICAL THEORY FOR THE *B* COEFFICIENT

Using Eq. (8.10) for A_{ki}, and Eq. (8.6) relating A_{ki} and B_{ki}, an expression is obtained for the stimulated emission coefficient:

$$B_{ki} = \frac{\mu_0}{4\pi} \frac{8\pi^3 c^2 e^2}{3h^2} \sum_{M_i} \langle k|\hat{\mathbf{r}}|i\rangle \cdot \langle i|\hat{\mathbf{r}}|k\rangle \tag{8.19}$$

and for the stimulated absorption coefficient:

$$B_{ik} = \frac{\mu_0}{4\pi} \frac{8\pi^3 c^2 e^2}{3h^2} \sum_{M_k} \langle k|\hat{\mathbf{r}}|i\rangle \cdot \langle i|\hat{\mathbf{r}}|k\rangle \tag{8.20}$$

These results would appear to be a factor of 2π less than that often quoted. It must be remembered, however, that the probability of transition from level k to level i is determined by $B_{ki}\rho(\nu_{ki})$ and, according to the note following Eq. (8.4), $\rho(\nu) = 2\pi\rho(\omega)$. The factor 2π is therefore recovered in the product $B\rho$ when transition probabilities are calculated.

The stimulated rates are determined by the products $B\rho$ as in Eq. (8.2), and, in order to establish relations between the A and B coefficients, ρ was taken to be the spectral distribution of the black-body radiation. This has the functional form given by Eq. (8.4), and is isotopic in angular distribution. Now the stimulated rate is not going to depend on the details of the complete spectral distribution but only on the energy density at the resonant frequency ν_{ki} of the transition. From this point of view we can replace $\rho(\nu_{ki})$ in the rate equations by $s(\nu_{ki})/c$, where $s(\nu_{ki})$ is the spectral intensity discussed in Chapter 5. However, when the radiation comes primarily from one direction, it is likely that the polarization of the radiation will limit the contribution in the sum over matrix elements in Eqs. (8.19) and (8.20). The polarization unit-vectors $\boldsymbol{\pi}$, which are implicit in expressions like Eq. (5.12) for the spectral density [and Eq. (4.23)] become transferred to the matrix elements of Eqs. (8.19) and (8.20) so that $\hat{\mathbf{r}}$ must be replaced by $\hat{\mathbf{r}} \cdot \boldsymbol{\pi}$. This modification is discussed in the next chapter.

CHAPTER 9

The Semiclassical Treatment of Stimulated Absorption and Emission

9.1. THE QUANTUM EQUATION OF MOTION

We are to study here the dynamical behavior of an atom, stimulated to make transitions between levels by interaction with an electromagnetic field. The discussion is restricted to two levels of the atom, a lower state $|i\rangle$ and an upper state $|k\rangle$, which are both regarded as nondegenerated; in the end the results can be expanded to the case of degenerate levels by summation over substates and by the inclusion of statistical weights.

We are concerned with transitions stimulated by the appropriate (near-resonant) frequency components of the field. The answer should give, in a more detailed way, what is described in the phenomenological approach using Einstein coefficients of Chapter 8, when the equilibrium is given by

$$N_k B_{ki} \rho(\nu_{ki}) + N_k A_{ki} = N_i B_{ik} \rho(\nu_{ki})$$

In the limit, when $\rho(\nu_{ki})$ is sufficiently large (i.e., for an intense enough stimulating field), the term in A_{ki} may be neglected. Thus, initially, we shall neglect entirely the spontaneous decay from $k \to i$, i.e., we assume $\gamma_k = 0$. The states $|i\rangle$ and $|k\rangle$ are regarded as stationary states of the Hamiltonian $\hat{H}_A$ for the internal energy of the atom:

$$\begin{aligned} \hat{H}_A|k\rangle &= W_k|k\rangle \\ \hat{H}_A|i\rangle &= W_i|i\rangle \end{aligned} \tag{9.1}$$

The *general* state of this two-level atom is described by a linear superposition of these two eigenstates of $\hat{H}_A$:

$$|t\rangle = a_k\, e^{-i2\pi W_k t/h}|k\rangle + a_i\, e^{-i2\pi W_i t/h}|i\rangle \tag{9.2}$$

where a_i, a_k are, in general, functions of time. We may choose $W_i = 0$ and write

$$\nu_0 = \nu_{ki} = \frac{W_k - W_i}{h} = \frac{W_k}{h}$$

for the *natural* frequency of the transition. Thus we have

$$|t\rangle = a_k\, e^{-i2\pi\nu_0 t}|k\rangle + a_i|i\rangle \tag{9.3}$$

When the electromagnetic field interacts with the atom, the total Hamiltonian, $\hat{H} = \hat{H}_A + \hat{H}_{\text{int}}$, includes the *interaction energy*:

$$\hat{H}_{\text{int}} = -\mathbf{E}(t)\cdot\hat{\mathbf{D}} = +e\mathbf{E}\cdot\hat{\mathbf{r}} \tag{9.4}$$

This is just the usual expression for the interaction energy between an electric field $\mathbf{E}$ and an electric dipole, $\mathbf{D} = -e\mathbf{r}$, except that, since the latter is a physical property of the atom, it must be expressed as an operator acting on the states of the atom.

The equation of motion, describing the dynamical behavior of the atom, uses Eq. (7.30) rather than the classical Eq. (3.2), or Eq. (3.25):

$$\frac{ih}{2\pi}\frac{\partial}{\partial t}|t\rangle = \hat{H}|t\rangle = (\hat{H}_A + \hat{H}_{\text{int}})|t\rangle \tag{9.5}$$

Placing Eq. (9.3) in this, we obtain

$$\frac{ih}{2\pi}\,\dot{a}_k\, e^{-i2\pi\nu_0 t}|k\rangle + \frac{ih}{2\pi}\,\dot{a}_i|i\rangle = -a_k\, e^{-i2\pi\nu_0 t}\mathbf{E}\cdot\hat{\mathbf{D}}|k\rangle - a_i\mathbf{E}\cdot\hat{\mathbf{D}}|i\rangle \tag{9.6}$$

The operator $\hat{\mathbf{D}}$ corresponds to a physical quantity of odd parity; thus the only matrix elements $\mathbf{D}_{ab} = -e\langle a|\hat{\mathbf{r}}|b\rangle$ that are nonzero are those between states $|a\rangle$ and $|b\rangle$ that are themselves of opposite parity (see Section 7.7). Using this selection rule, and multiplying Eq. (9.6) through by $(2\pi/ih)\langle k|\, e^{+i2\pi\nu_0 t}$, and again through by $(2\pi/ih)\langle i|$, we obtain the coupled equations

$$\begin{aligned}\dot{a}_k &= \frac{i2\pi}{h}\mathbf{E}\cdot\langle k|\hat{\mathbf{D}}|i\rangle\, e^{+i2\pi\nu_0 t}a_i\\ \dot{a}_i &= \frac{i2\pi}{h}\mathbf{E}\cdot\langle i|\hat{\mathbf{D}}|k\rangle\, e^{-i2\pi\nu_0 t}a_k\end{aligned} \tag{9.7}$$

$\mathbf{E} = \mathbf{E}(t)$ is the electric vector of the field which, in general, has a spectrum of frequencies and can be expanded according to the Fourier transformation Eq. (3.16).

9.2. STIMULATION BY MONOCHROMATIC RADIATION

First, we shall consider the monochromatic case, radiation of frequency ν given by Eq. (3.5):

$$\mathbf{E}(t) = \frac{E_0}{\sqrt{2}}[\boldsymbol{\pi}^+ e^{-i2\pi\nu t} + \boldsymbol{\pi}^- e^{+i2\pi\nu t}] \tag{9.8}$$

In the absence of the electromagnetic radiation the atom will rest in the lower level, $|a_i| = 1$, $a_k = 0$. Significant transfer of population to the excited state $|k\rangle$ will occur only when the stimulating radiation contains frequencies ν close to ν_0; therefore, when substituting for $\mathbf{E}(t)$ in Eq. (9.7), we shall retain only "near-resonant" terms in $\nu - \nu_0$ rejecting "antiresonant" terms in $\nu + \nu_0$ (in the integration of the differential equation the term would become small). This is frequently referred to as the *rotating wave approximation.* Thus Eq. (9.7) can be written

$$\begin{aligned} \dot{a}_k &= \frac{i\sqrt{2}\,\pi}{h} E_0 \boldsymbol{\pi}^+ \cdot \mathbf{D}_{ki}\, e^{-i2\pi(\nu-\nu_0)t} a_i = i\pi\nu_R\, e^{-i2\pi\delta t} a_i \\ \dot{a}_i &= \frac{i\sqrt{2}\,\pi}{h} E_0 \boldsymbol{\pi}^- \cdot \mathbf{D}_{ik}\, e^{+i2\pi(\nu-\nu_0)t} a_k = i\pi\nu_R\, e^{+i2\pi\delta t} a_k \end{aligned} \tag{9.9}$$

The difference frequency, δ has been defined as

$$\delta = \nu - \nu_0 \tag{9.10}$$

The frequency ν_R is called the Rabi frequency:

$$\nu_R = \frac{\sqrt{2}\, E_0(\boldsymbol{\pi}^+ \cdot \mathbf{D}_{ki})}{h} = \frac{\sqrt{2}\, E_0(\boldsymbol{\pi}^- \cdot \mathbf{D}_{ik})}{h} \tag{9.11}$$

It is related to the intensity of the stimulating radiation by

$$\begin{aligned} \nu_R^2 &= \frac{2E_0^2(\boldsymbol{\pi}^+ \cdot \mathbf{D}_{ki})(\boldsymbol{\pi}^- \cdot \mathbf{D}_{ik})}{h^2} \\ &= \frac{2\mu_0 c I}{h^2}(\boldsymbol{\pi}^+ \cdot \mathbf{D}_{ki})(\boldsymbol{\pi}^- \cdot \mathbf{D}_{ik}) \end{aligned} \tag{9.12}$$

where I is the intensity of the radiation; in this case $I = E_0^2/\mu_0 c$; see Eq. (4.12).

Eliminating a_i between the two equations of Eq. (9.9), a second-order differential equation for a_k is obtained:

$$\ddot{a}_k + i2\pi\delta\dot{a}_k + \pi^2\nu_R^2 a_k = 0 \tag{9.13}$$

The equation is readily solved by introducing throughout an integrating factor $e^{i\pi\delta t}$; the differential equation can then be written as (using $D^2 = \partial^2/\partial t^2$):

$$[D^2 + (\pi\nu'_R)^2]a_k e^{i\pi\delta t} = 0 \tag{9.14}$$

where ν'_R is the *modified Rabi frequency*:

$$\nu'_R = (\nu_R^2 + \delta^2)^{1/2} \tag{9.15}$$

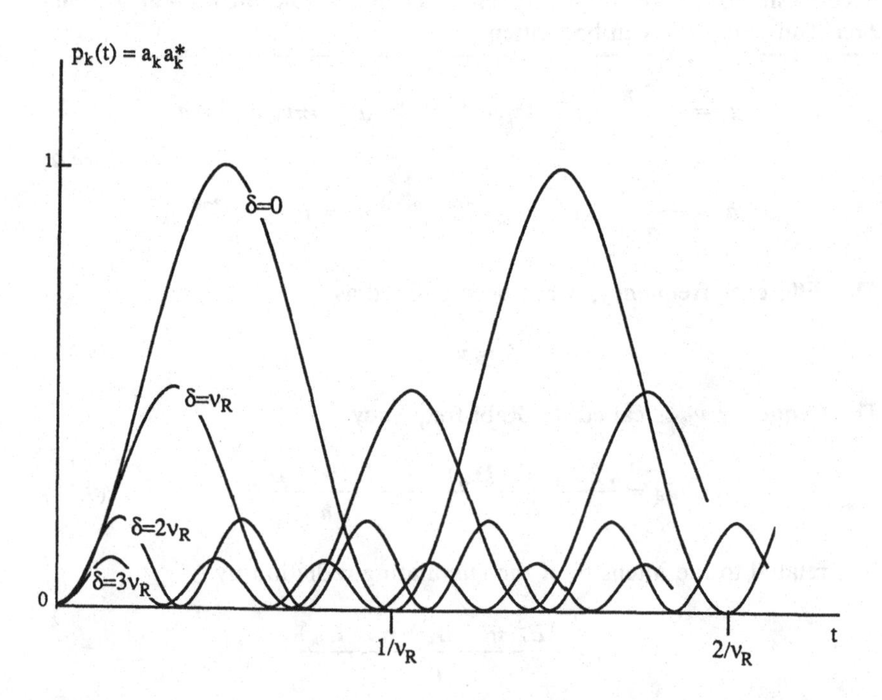

FIGURE 10. Stimulated excitation of an excited state; the probability $p_k(t)$ oscillates at the Rabi frequency. The difference frequency $\delta = \nu - \nu_0$ is that between the frequency of monochromatic stimulation and the transition frequency.

Equation (9.14) has the solution

$$a_k = e^{-i\pi\delta t}[A\sin(\pi\nu'_R t) + B\cos(\pi\nu'_R t)]$$

$$= F e^{-i\pi(\delta-\nu'_R)t} + G e^{-i\pi(\delta+\nu'_R)t} \tag{9.16}$$

with the constants A and B (or F and G) still to be determined according to the initial conditions. Under the initial condition $a_k = 0$, $a_i = 1$ at $t = 0$, Eqs. (9.9) and (9.16) can be solved to give $A = i(\nu_R/\nu'_R)$, $B = 0$ (or $F = -G = \nu_R/2\nu'_R$):

$$a_k = i\frac{\nu_R}{\nu'_R} e^{-i\pi\delta t}\sin(\pi\nu'_R t)$$

$$a_i = -i\frac{\delta}{\nu'_R} e^{+i\pi\delta t}\sin(\pi\nu'_R t) + e^{+i\pi\delta t}\cos(\pi\nu'_R) \tag{9.17}$$

It can be readily checked that $a_k a_k^* + a_i a_i^* = 1$, as required by the normalization of the state $|t\rangle$ given by Eq. (9.3).

The probability, $p_k(t)$, of finding the atom in the upper state $|k\rangle$ at time t is

$$p_k(t) = a_k a_k^*$$

$$= \left(\frac{\nu_R}{\nu'_R}\right)^2 \sin^2(\pi\nu'_R t)$$

$$= \left(\frac{\nu_R}{\nu'_R}\right)^2 \tfrac{1}{2}[1 - \cos(2\pi\nu'_R t)] \tag{9.18}$$

The meaning of the Rabi frequency is now clear: the probability that the stimulated atom will be found in the excited state oscillates with a frequency ν'_R (or ν_R if the stimulating radiation is on resonance with the atomic transition). The behavior is illustrated in Fig. 10.

9.3. STIMULATION BY POLYCHROMATIC RADIATION

The generalization of these results to the case where the field is polychromatic rather than monochromatic is not straightforward. We shall adopt a simple approach. Note first that the Rabi frequency ν_R defined by Eq. (9.11) or Eq. (9.12) depends only on the intensity of the monochromatic radiation and not on its frequency; the only restriction in the analysis that leads to this result is that the "neoresonant" term, $(\nu - \nu_0)$, has been retained and the "antiresonant" term, $(\nu + \nu_0)$, has been rejected. We have thereby made the assumption that the frequency of the field is not too remote from the natural transition frequency. The point is emphasized by the final result for $a_k a_k^*$, which, according to Eq. (9.18) and Fig. 10, would never achieve a value significantly different from zero unless ν were reasonably close to ν_0.

Thus, in order to generalize the result, we shall regard Eq. (9.17) to be valid for the polychromatic case with the variation according to applied frequency ν contained in the factor $\nu'_R = (\nu_R^2 + \delta^2)^{1/2}$. Then, in forming the product $a_k a_k^*$, one must carry out the averaging process over optical periods and frequencies as is done in Appendix 1. This approximation is equivalent to averaging Eq. (9.18) over the spectral intensity distribution $s(\nu)$ defined in Eqs. (A1.18) and (A1.19). The result for $p_k(t)$ is

$$p_k(t) = a_k a_k^* = \int_0^\infty d\nu \left(\frac{\nu_R}{\nu'_R}\right)^2 \frac{s(\nu)}{I} \sin^2(\pi \nu'_R t) \tag{9.19}$$

In carrying out the averaging over time, the averaging period T must be restricted in order to retain the Rabi frequency ν_R; the method for doing this is discussed in Appendix 1. There are some contentious points in the above analysis; its validity rests partly on the success of Eq. (9.19) in predicting acceptable results.

9.4. DEDUCTION OF THE EINSTEIN B COEFFICIENT

The results of the last section can be applied to deduce an expression for the Einstein B coefficient. It must be remembered that, in Einstein's treatment of the stimulated process (Chapter 8), the following was assumed:

i. That the radiation was broad-band rather than monochromatic. Therefore Eq. (9.19) is to be evaluated with $s(\nu)$ presumed reasonably constant over a frequency band centered on ν_0 and of width several times the Rabi frequency ν_R.
ii. That the radiation was isotropic. Therefore one must average over all possible polarizations.

Regarding $s(\nu)$ as being constant and equal to $s(\nu_0)$ over the range of frequencies where the integrand makes its major contributions (that is, where $\nu \approx \nu_0$), Eq. (9.19) is written

$$p_k(t) = \frac{\nu_R s(\nu_0)}{I} \int_{-\infty}^{\infty} dx \frac{\sin^2[\pi \nu_R (1 + x^2)^{1/2} t]}{(1 + x^2)} \tag{9.20}$$

where $x = \delta/\nu_R = (\nu - \nu_0)/\nu_R$; the lower limit of the integral has been placed equal to $-\infty$, this being a good approximation to $-\nu_0/\nu_R$. Since the integral is an even function of x, it may be expressed as twice the integral over positive values of x only. The rate at which the state $|k\rangle$ becomes populated is

$$\frac{dp_k(t)}{dt} = \frac{2\pi\nu_R^2 s(\nu_0)}{I} \int_0^\infty dx \frac{\sin[2\pi\nu_R t(1+x^2)^{1/2}]}{(1+x^2)^{1/2}}$$

$$= \frac{\pi^2\nu_R^2 s(\nu_0)}{I} J_0(2\pi\nu_R t) \tag{9.21}$$

using a standard integration [Gradsteyn and Ryzhik, Ref. 1, Eq. (3.876.1)]. Evaluating this at $t = 0$ [the Bessel function $J_0(0) = 1$], when the rate of increase of population of state $|k\rangle$ is entirely due to stimulation from state $|i\rangle$, we obtain

$$\left(\frac{dp_k(t)}{dt}\right)_{t=0} = \frac{\pi^2\nu_R^2 s(\nu_0)}{I}$$

$$= \frac{2\pi^2\mu_0 c}{h^2}(\boldsymbol{\pi}^+ \cdot \mathbf{D}_{ki})(\boldsymbol{\pi}^- \cdot \mathbf{D}_{ik})s(\nu_0) \tag{9.22}$$

where Eq. (9.12) has been used. The average over all polarization directions must now be taken. This is obviously independent of whether the polarization is linear, elliptical, or circular and can most readily be evaluated using linearly polarized light. Using the notation $\langle\ \rangle_\pi$ to mean "average over all directions of the polarization vector, $\cos\beta = \boldsymbol{\pi} \cdot \mathbf{D}$":

$$\langle(\boldsymbol{\pi}^+ \cdot \mathbf{D}_{ki})(\boldsymbol{\pi}^- \cdot \mathbf{D}_{ik})\rangle_\pi = \mathbf{D}_{ki} \cdot \mathbf{D}_{ik}\langle\cos^2\beta\rangle_\pi = \frac{\mathbf{D}_{ki} \cdot \mathbf{D}_{ik}}{3} \tag{9.23}$$

Substituting in Eq. (9.22), we obtain a quantum-mechanical expression for the rate at which state $|k\rangle$ is populated by transitions from state $|i\rangle$, caused by isotropic radiation of spectral intensity $s(\nu_0)$:

$$\left\langle\left(\frac{dp_k(t)}{dt}\right)_{t=0}\right\rangle_\pi = \frac{2\pi^2\mu_0 c}{3h^2}\mathbf{D}_{ki} \cdot \mathbf{D}_{ik} s(\nu_0)$$

This, by the definition given in Eq. (8.2), equals $B_{ik}\rho(\nu_0) = B_{ik}s(\nu_0)/c$. Hence we have

$$B_{ik} = \frac{\mu_0}{4\pi}\frac{8\pi^3 c^2 e^2}{3h^2}\langle k|\hat{\mathbf{r}}|i\rangle \cdot \langle i|\hat{\mathbf{r}}|k\rangle \tag{9.24}$$

The result has been deduced using nondegenerate states $|i\rangle$ and $|k\rangle$. After summing over degenerate substates and introducing statistical weight the result is exactly that deduced by Einstein, Eq. (8.20)

$$B_{ik} = \frac{\mu_0}{4\pi} \frac{8\pi^3 c^2 e^2}{3h^2} \sum_{M_k} \langle k|\hat{\mathbf{r}}|i\rangle \cdot \langle i|\hat{\mathbf{r}}|k\rangle \tag{9.25}$$

We have shown that the theory of stimulated transition using a semiclassical perturbation approach leads to the same result as does the phenomenological approach to Einstein.

REFERENCE

1. I. S. GRADSTEYN AND L. M. RYZHIK, *Table of Integrals, Series, and Products* (Academic, New York, 1965).

CHAPTER 10

THE SEMICLASSICAL DESCRIPTION OF SPONTANEOUS DECAY

10.1. THE ONE-WAY NATURE OF DECAY

Formulas for the decay constant have been deduced in two places above. In the classical approach, the decay constant is defined through the equation

$$\gamma = \frac{-(dW/dt)}{W} = \frac{P(t)}{W(t)} = \frac{P(0)}{W(0)}$$

where $W(t)$ is the energy of the system at time t, and $P(t)$ is the power of dissipation (or radiation). Classical electromagnetic theory gives Eq. (3.31) for $P(0)$, and the classical energy of a mechanical oscillator is $W(0) = 4\pi^2\nu_0^2 mA_0^2$. The result for the classical decay constant of an oscillating and radiating charge is Eq. (3.27):

$$\gamma_{\text{class}} = \frac{\mu_0}{4\pi}\frac{8\pi^2 e^2}{3mc}\nu_0^2 \tag{10.1}$$

Alternatively, Einstein's theory for the spontaneous decay rate between two levels of an atom, k and i, uses $W = h\nu_0$ for the energy stored in the excited system, and leads to Eq. (8.10) for the spontaneous decay rate from level k to a lower level i. From this the total decay constant of level k to all lower levels i is obtained:

$$\begin{aligned}\gamma_k &= \sum_{i<k} A_{ki} \\ &= \sum_{i<k} \frac{\mu_0}{4\pi}\frac{64\pi^4 e^2 \nu_{ki}^3}{3ch} \sum_{M_i} \langle J_k M_k|\hat{\mathbf{r}}|J_i M_i\rangle \cdot \langle J_i M_i|\hat{\mathbf{r}}|J_k M_k\rangle\end{aligned} \tag{10.2}$$

In this chapter we seek a dynamical theory of the atom that leads to this result, rather than through the phenomenological approach of Einstein.

The first thing to remark on is that, as observed in nature, spontaneous transition occurs only in one direction; an excited level will decay to a lower one provided certain selection rules are satisfied, but the reverse transition (spontaneous absorption) does not occur. It is, philosophically, the same problem as the mechanical oscillator which can dissipate its energy through friction to a heat bath, but can never absorb it again to regain its mechanical energy. The second law of thermodynamics interprets this fact as a law of nature, and injects an "arrow of time" into physical behavior by introducing the concept of entropy. The equations of motion in quantum mechanics (Schrödinger's equation, for example), like those of classical mechanics without the second law of thermodynamics, are symmetric in time, a fact that leads to a common theory for stimulated absorption and emission as in Chapter 9. For a theory of spontaneous emission one therefore expects some extra feature to be introduced into the theory, a feature that injects an "arrow of time" into the quantum mechanical description. The physical reason for the one-way nature of the radiation process is readily appreciated; the emitted radiation disappears into the infinite bath of space from which it never again returns to be reabsorbed by the atom.

However, it is necessary to have some physical process by which spontaneous emission is initiated and maintained. It would be expected that an *isolated* atom, described by a Hamiltonian $\hat{H}_A$ (the atomic Hamiltonian) representing the energy of the *internal* interaction, would be in a *stationary* state $|j\rangle$ one of the eigenstates of $\hat{H}_A$ with an energy $W_j = h\nu_j$. And there it would stay. To achieve a transition some perturbation is required, represented by an addition to the Hamiltonian, $\hat{H}_D$, the decay Hamiltonian. In some way the action of $\hat{H}_D$ must, either through its intrinsic nature or through some restriction placed on its operation, give decay only and not absorption. It must also lead to the result of Eq. (10.2). Various possibilities are now discussed.

Firstly, restricting our discussion to a two-state system, the atom is radiating at a frequency $\nu_0 = \nu_k - \nu_i$. According to the classical argument, the radiation reacts on the atom with a *radiation reaction* force as discussed in Chapter 3. In interaction with the state of the atom this causes a perturbation described by $\hat{H}_D = \hat{H}_{RR}$, the radiation reaction Hamiltonian. This is to be discussed in Section 10.2.

A second possibility is to postulate the existence of a background electromagnetic field, the classical equivalent of the zero-point fluctuation of quantum field theory. This could also be thought of as a kind of thermodynamic equilibrium with all the atoms of the universe. The radiation from an excited atom is ultimately absorbed by an atom or molecule in the universe causing an upward transition of some atom in its ground state. In their turn the remote absorbing atoms of the universe themselves contribute to a bath of background radiation which cause downward transitions of excited atoms. The effect of *zero-point radiation* is described by a perturbation $\hat{H}_D = \hat{H}_{ZP}$, the zero-point Hamiltonian. This is to be discussed in Section 10.3 (see, e.g., D. T. Pegg[1]).

We proceed to discuss spontaneous decay from these two points of view. As mentioned above, when using a perturbation theory we shall inject rules to ensure the one-way nature of the process; in doing so we can derive some comfort from the fact that a similar thing is done in classical mechanics with the introduction of the second law of thermodynamics, and that a quantum electrodynamical theory does something equivalent when *normal ordering* of operators is adopted. The aim is to achieve a recipe that works for a wide range of examples.

10.2. SPONTANEOUS EMISSION AS DERIVED FROM RADIATION REACTION

Consider the spontaneous decay of an atom initially, i.e., at $t=0$, in a state $|k\rangle$. The general state of the atom at later time t is written as

$$|t\rangle = a_k\, e^{-i2\pi\nu_k t}|k\rangle + \sum_j a_j\, e^{-i2\pi\nu_j t}|j\rangle \tag{10.3}$$

where the amplitudes $a_k = a_k(t)$ and $a_j = a_j(t)$ are slowly varying functions of time. Initially, because the atom is prepared in the excited state, $a_k^*(0)a_k(0) = 1$, and all other $a_j(0) = 0$. The summation is now restricted to those states $|i\rangle$ lower in energy than state $|k\rangle$, which are accessible by electric dipole transition:

$$|t\rangle = a_k\, e^{-i2\pi\nu_k t}|k\rangle + \sum_i a_i\, e^{-i2\pi\nu_i t}|i\rangle \tag{10.4}$$

Although, for the two-level atom, the lower level may be restricted to one, the summation is retained for two reasons:

i. The lower level may be degenerate (or near degenerate) and the summation then runs over all degenerate labels.
ii. The radiation is emitted in all directions in space. Thus the final state of the system is of the atom in the ground state $|i\rangle$, plus radiation propagating in a large number of possible modes. As we saw when studying the Compton effect in Chapter 6, the atom recoils to take up the opposite momentum. Thus the final state of the atom is one of a large number (infinite) of possible states, each of the same eigenstate $|i\rangle$ of $\hat{H}_A$, but with a distribution of linear momenta in the laboratory.

The Hamiltonian of the system is now

$$\hat{H} = \hat{H}_A + \hat{H}_{RR} \tag{10.5}$$

where $\hat{H}_A$ has eigenstates defined by

$$\begin{aligned} \hat{H}_A|k\rangle &= W_k|k\rangle = h\nu_k|k\rangle \\ \hat{H}_A|i\rangle &= W_i|i\rangle = h\nu_i|i\rangle \end{aligned} \tag{10.6}$$

The electric field vector that creates the radiation-reaction force that acts on the atomic electron is, using Eq. (3.23),

$$\mathbf{E}_{RR} = \frac{\mathbf{F}_{RR}}{-e} = -\frac{m\tau}{e}\dddot{\mathbf{x}} \tag{10.7}$$

When this interacts with an atomic dipole moment, $\mathbf{D} = -e\mathbf{x}$, the interaction energy is

$$-\mathbf{D}\cdot\mathbf{E}_{RR} = -m\tau\mathbf{x}\cdot\dddot{\mathbf{x}} \tag{10.8}$$

The perturbing Hamiltonian operator can be written as

$$\hat{H}_{RR} = -m\tau\hat{\mathbf{x}}\cdot\hat{\partial}^3\hat{\mathbf{x}} \tag{10.9}$$

where we write $\hat{\partial}$ for the time differential operator $\partial/\partial t$. The triple time differentiation is an operation that will act on the temporal factors in any state vector placed on its right. Hence the Hamiltonian is written

$$\hat{H}_{RR} = -m\tau\hat{\mathbf{x}}\cdot\hat{\mathbf{x}}\,\hat{\partial}^3 \tag{10.10}$$

Within the eigenstates of the expansion Eq. (10.4), this has only diagonal matrix elements.

The equation of motion

$$\frac{ih}{2\pi}\frac{\partial}{\partial t}|t\rangle = \hat{H}|t\rangle = (\hat{H}_A + \hat{H}_{RR})|t\rangle$$

can be expanded, using Eqs. (10.4) and (10.6), as

$$\begin{aligned} &\frac{ih}{2\pi}\dot{a}_k|k\rangle\,e^{-i2\pi\nu_k t} + \sum_i \frac{ih}{2\pi}\dot{a}_i|i\rangle\,e^{-i2\pi\nu_i t} \\ &\qquad = -m\tau\hat{\mathbf{x}}\cdot\hat{\mathbf{x}}\,\hat{\partial}^3 a_k|k\rangle\,e^{-i2\pi\nu_k t} - m\tau\sum_i \hat{\mathbf{x}}\cdot\hat{\mathbf{x}}\,\hat{\partial}^3 a_i|i\rangle\,e^{-i2\pi\nu_i t} \end{aligned} \tag{10.11}$$

Premultiply through by $(2\pi/ih)\, e^{+i2\pi\nu_k t}\langle k|$:

$$\dot{a}_k = \frac{i2\pi m\tau}{h}\, e^{+i2\pi\nu_k t}\langle k|\hat{\mathbf{x}} \cdot \hat{\mathbf{x}}|k\rangle \hat{\partial}^3 a_k\, e^{-i2\pi\nu_k t} \tag{10.12}$$

Now enter the unit dyadic operator, a modification of Eq. (7.9):

$$\mathbf{I} = \sum_j e^{-i2\pi\nu_j t}|j\rangle\langle j|\, e^{+i2\pi\nu_j t} \tag{10.13}$$

the summation being over all states of the system. The only terms that survive are those that have nonzero matrix elements $\langle j|\hat{\mathbf{x}}|k\rangle$. These include states that are both higher and lower than $|k\rangle$ in energy; we shall include only those states $|i\rangle$ of lower energy. This is the point at which the "arrow of time" is injected. The physical justification for the mathematical restriction is that, for states $|j\rangle$ of higher energy than $|k\rangle$, no source of energy exists for an atom in empty space to make that transition. However, for states $|i\rangle$ of lower energy, the energy can be radiated into the reservoir of the surrounding space, eventually perhaps to be absorbed by (remote) atoms of the universe. Then we have

$$\dot{a}_k = \frac{i2\pi m\tau}{h} \sum_i e^{+i2\pi\nu_0 t}\langle k|\hat{\mathbf{x}}|i\rangle \cdot \langle i|\hat{\mathbf{x}}|k\rangle \hat{\partial}^3\, e^{-i2\pi\nu_0 t} a_k \tag{10.14}$$

with $\nu_0 = \nu_k - \nu_i$. To a first approximation we may regard a_k as changing only slowly with time and ignore its time differential on the right. After carrying out the differentiation and entering the expression for τ from Eq. (3.22) we have

$$\dot{a}_k = -\sum_i \frac{\mu_0}{4\pi} \frac{32\pi^4 e^2 \nu_0^3}{3hc} \langle k|\hat{\mathbf{x}}|i\rangle \cdot \langle i|\hat{\mathbf{x}}|k\rangle a_k \tag{10.15}$$

The probability of finding the atom in state $|k\rangle$ at time t is $p_k(t) = a_k^* a_k$. The rate of decay of this is

$$\dot{p}_k(t) = \dot{a}_k^* a_k + a_k^* \dot{a}_k$$

Using Eq. (10.15) and its complex conjugate, we obtain

$$\dot{p}_k(t) = -\sum_i \frac{\mu_0}{4\pi} \frac{64\pi^4 e^2 \nu_0^3}{3hc} \langle k|\hat{\mathbf{x}}|i\rangle \cdot \langle i|\hat{\mathbf{x}}|k\rangle p_k(t)$$

with the solution

$$p_k(t) = p_k(0)\, e^{-\gamma_k t}$$

in which

$$\gamma_k = \sum_i \frac{\mu_0}{4\pi} \frac{64\pi^4 e^2 \nu_0^3}{3hc} \langle k|\hat{\mathbf{x}}|i\rangle \cdot \langle i|\hat{\mathbf{x}}|k\rangle = \sum_i A_{ki} \tag{10.16}$$

where A_{ki} is the Einstein A coefficient as deduced in Eq. (8.9).

We have deduced the expected result, but we must entertain some suspicions about the validity of certain steps in the analysis. The analysis may be felt to have a physical justification but to lack mathematical rigidity.

Between Eq. (10.14) and Eq. (10.15) certain approximations were adopted. When the analysis is carried to the next order it results in a shift of the energy difference between the levels involved from $h\nu_0$, to $h(\nu_0 + \delta)$. The frequency shift induced by radiation reaction is $\delta = 3\gamma^2/16\pi^2\nu_0$, which, for the Lyman-$\alpha$ line discussed in Section 8.3, has the extremely small value of $\approx$10 Hz, a figure that justifies its neglect.

10.3. SPONTANEOUS EMISSION AS DERIVED FROM THE ZERO-POINT VACUUM FIELD

If an electromagnetic field were to exist in zero temperature vacuum its properties must be consistent with that state. A vacuum is homogeneous, i.e., the same at all points; therefore the zero-point field must be independent of spatial coordinates. A vacuum is isotropic, i.e., has no preferred direction; therefore the zero-point field must have a uniform angular distribution both in propagation and polarization directions. Furthermore, in a vacuum all inertial frames are equivalent; therefore the zero-point field must be Lorentz invariant, it must look the same to all observers in uniform relative motion. It is shown in Appendix 3 that a zero-point field with these properties is given by

$$\begin{aligned}\mathbf{E}_{\text{ZP}}(t) &= \frac{b}{2}\sum_\lambda \int_0^\infty d\nu\, \nu \int_{-1}^{+1} d(\cos\theta) \int_0^{2\pi} d\phi\, \boldsymbol{\pi}^0(\nu\mathbf{k}^0\lambda) \\ &\quad \times [e^{-i2\pi\nu t}\, e^{i\varphi(\nu\mathbf{k}^0\lambda)} + e^{+i2\pi\nu t}\, e^{-i\varphi(\nu\mathbf{k}^0\lambda)}] \\ &= \mathbf{E}^+_{\text{ZP}}(t) + \mathbf{E}^-_{\text{ZP}}(t)\end{aligned} \tag{10.17}$$

In Eq. (10.17) $\mathbf{k}^0$ is the unit vector, with polar angle θ and azimuthal angle ϕ, defining a direction of propagation; $\boldsymbol{\pi}^0$ is a unit vector defining polarization and is, in general, a random function of ν and $\mathbf{k}^0$ but with the orthogonal property $\boldsymbol{\pi}^0 \cdot \mathbf{k}^0 = 0$; λ labels a summation over two independent states of polarization with the orthogonal property: $\boldsymbol{\pi}^0(\lambda) \cdot \boldsymbol{\pi}^0(\lambda_1) = \delta(\lambda\lambda_1)$; φ is the phase, again a random function of ν, $\mathbf{k}^0$, and λ.

The factor b is an invariant scalar physical quantity defining the amplitude; its value is yet to be determined. The field $\mathbf{E}_{ZP}(t)$ has been broken into two parts, $\mathbf{E}^+_{ZP}(t)$ associated with the factor $e^{-i2\pi\nu t}$ and its complex conjugate $\mathbf{E}^-_{ZP}(t)$.

The existence of such a zero-point field has been invoked (see Boyer[(2)]) to explain the Casimir effect, i.e., the attraction of two conducting plates suspended close to each other in vacuum. Experimental determination of this small force is consistent with

$$b^2 = \frac{\mu_0}{4\pi}\frac{6h}{c} \tag{10.18}$$

As shown in Appendix 3, this value for b endows the zero-point field with an energy of $h\nu/2$ per normal mode. It is shown below that this value also gives the correct value for the rate of spontaneous decay of an excited atom.

This zero-point field has a spectral energy density of [Eq. (A3.13)]

$$\rho^+_{ZP}(\nu) = \frac{4\pi h\nu^3}{c^3} = \frac{1}{2}\left(\frac{8\pi\nu^2}{c^3}\right)h\nu \tag{10.19}$$

The ν^3 dependence is a requirement for a field that obeys Lorentz invariance as is demonstrated in Appendix 3; *an isotropic electromagnetic field with energy density proportional to ν^3 is uniquely Lorentz invariant.*

It is also essential that this field shall not stimulate *absorptive* transition in an atom placed in the vacuum; it is for this reason that the field $\mathbf{E}_{ZP}$ has been broken, in Eq. (10.17), into two analytic terms, $\mathbf{E}^+_{ZP}(t)$ and $\mathbf{E}^-_{ZP}(t)$. We shall find that only the bilinear product in the order $\mathbf{E}^+_{ZP}(t) \cdot \mathbf{E}^-_{ZP}(t)$ participates in the interaction, providing the required "arrow of time"; thus it is the "half-field" only, i.e., Eq. (10.17) with the omission of the complex-conjugate term, which forms the correct expression for the zero-point field. This is also the reason for the $^+$ superscript on ρ^+_{ZP} in Eq. (10.19).

The Hamiltonian describing the perturbation of $\mathbf{E}_{ZP}$ on the atom is

$$\hat{H}_{ZP} = -\mathbf{E}_{ZP}(t) \cdot \hat{\mathbf{D}}$$

The equation of motion of the system is

$$\frac{ih}{2\pi}\frac{\partial}{\partial t}|t\rangle = \hat{H}|t\rangle$$

with $\hat{H} = \hat{H}_A + \hat{H}_{ZP}$, and the atomic Hamiltonian $\hat{H}_A$ having eigenstates defined in Eq. (10.6).

The general state of the atom at time t is written as

$$|t\rangle = a_k(t)\, e^{-i2\pi\nu_k t}|k\rangle + a_i(t)\, e^{-i2\pi\nu_i' t}|i\rangle \tag{10.20}$$

It is to be recognized that $|i\rangle$ here is not a single state as discussed after Eq. (10.4). When a stationary atom in an excited state $|k\rangle$, of total energy W_k, decays by spontaneous emission to a ground state whose rest energy is W_i, the atom recoils with momentum equal and opposite to that of the emitted radiation; the states, energies, and momenta are illustrated in Fig. 11.

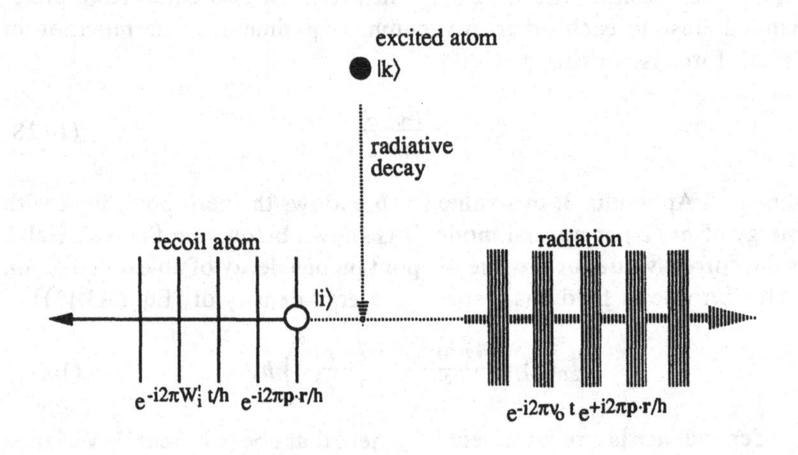

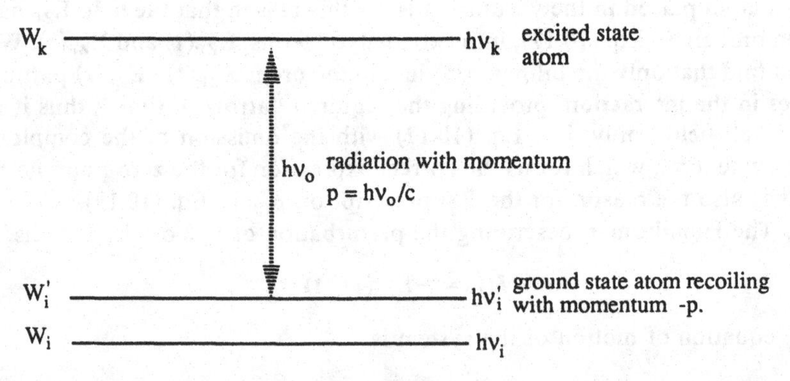

FIGURE 11. The states, energies, and momenta of an atom recoiling after the emission of radiation.

The energy of the emission cannot be the full amount given by $h\nu_{ki} = W_k - W_i$ because some energy is needed for the kinetic energy of recoil. If the frequency of the emitted radiation is ν_0, then the momenta of the radiation and of the recoil atom are given by

$$cp = h\nu_0 = W_k - W_i' = h\nu_k - h\nu_i'$$

The total energy of the recoiling ground-state atom, W_i', is thereby greater than its rest energy according to

$$h\nu_i' = W_i' = (W_i^2 + c^2p^2)^{1/2}$$

These last three equations provide the reason for the prime placed on ν_i' in Eq. (10.20). Strictly speaking therefore the second term on the right-hand side of Eq. (10.20) is an isotropic distribution of recoil states:

$$a_i(t)\, e^{-i2\pi\nu_i' t}|i\rangle$$
$$= \int_{-1}^{+1} d(\cos\theta_p) \int_0^{2\pi} d\phi_p\, b(t, \mathbf{p})\, e^{-i2\pi\nu_i' t}\, e^{-i2\pi \mathbf{p}\cdot\mathbf{r}/h}|i, \mathbf{p}\rangle \tag{10.21}$$

all of the same modulus of momentum. The angles θ_p and ϕ_p specify the angle of recoil of the atom with momentum $\mathbf{p}$. It is this infinite reservoir of final states that makes the transition irreversible. We simplify notation by writing

$$a_i(t)\, e^{-i2\pi\nu_i' t}|i\rangle = \int d\alpha\, b_i(t, \alpha)\, e^{-i2\pi\nu_i' t}|i\alpha\rangle \tag{10.22}$$

where α represents the orientation of $\mathbf{p}$, and the factor $e^{-i2\pi\mathbf{p}\cdot\mathbf{r}/h}$ can either be regarded as incorporated into the state vector $|i\alpha\rangle$, or has been placed equal to unity on the grounds that we can place $\mathbf{r} = 0$ because of the homogeneous nature of the system. Thus, after removal of an irrelevant phase, the state of the atom becomes

$$|t\rangle = a_k(t)\, e^{-i2\pi\nu_0 t}|k\rangle + \int d\alpha\, b_i(t, \alpha)|i\alpha\rangle \tag{10.23}$$

We have an orthogonality condition on the $|i\alpha\rangle$ states:

$$\langle i\alpha\,|\,i\alpha_1\rangle = \delta(\alpha\,\alpha_1) \tag{10.24}$$

There is also a closure condition:

$$|k\rangle\langle k| + |i\rangle\langle i| = 1 \tag{10.25}$$

where we have used the short-hand notation $\int d\alpha\, |i\alpha\rangle\langle i\alpha| = |i\rangle\langle i|$. Solving the equation of motion in the normal way we arrive at the coupled differential equations:

$$\dot{a}_k(t) = -\frac{2\pi}{ih}\int d\alpha\, b_i(t,\alpha)\mathbf{E}_{ZP}(t)\cdot\langle k|\hat{\mathbf{D}}|i\alpha\rangle\, e^{+i2\pi\nu_0 t} \tag{10.26}$$

$$\dot{b}_i(t,\alpha) = -\frac{2\pi}{ih}\, a_k(t)\mathbf{E}_{ZP}(t)\cdot\langle i\alpha|\hat{\mathbf{D}}|k\rangle\, e^{-i2\pi\nu_0 t} \tag{10.27}$$

Note the irreversibility of these equations; the first incorporates an integration over all final states.

The initial conditions for the system are $a_k(0) = 1$ and $b_i(0, \alpha) = 0$. Integrate Eq. (10.27) from $t' = 0$ to $t' = t$:

$$b_i(t,\alpha) = -\frac{2\pi}{ih}\int_0^t dt'\, a_k(t')\mathbf{E}_{ZP}(t')\cdot\langle i\alpha|\hat{\mathbf{D}}|k\rangle\, e^{-i2\pi\nu_0 t'} \tag{10.28}$$

Substitute in Eq. (10.26), and use the closure relation on $|i\alpha\rangle$, Eq. (10.25):

$$\dot{a}_k(t) = -\frac{4\pi^2}{h^2}\int_0^t dt'\, a_k(t')\, e^{+i2\pi\nu_0(t-t')}\mathbf{E}_{ZP}(t)\cdot\mathbf{D}_{ki}\mathbf{D}_{ik}\cdot\mathbf{E}_{ZP}(t') \tag{10.29}$$

The equation of motion has been reduced to a single integrodifferential equation in which $\dot{a}_k(t)$ depends on the entire history of the interaction between the field and the atom from $0 < t' < t$.

At this point we introduce the expansion for the zero-point field given by Eq. (10.17). The amplitudes $a_k(t)$ and $b_i(t, \alpha)$ are essentially quantities that vary slowly with time. Hence, in the integration over time variables, and in the ensemble averaging that will be used, it is only the cross product of the "half-fields" of Eq. (10.17), i.e., $\mathbf{E}^+_{ZP}(t)\cdot\mathbf{E}^-_{ZP}(t')$, which survives from the bilinear product of fields in Eq. (10.29)—the products like $\mathbf{E}^-\cdot\mathbf{E}^-$ and $\mathbf{E}^+\cdot\mathbf{E}^+$ have fast oscillating terms that must be discarded, and the cross product with ordering $\mathbf{E}^-\cdot\mathbf{E}^+$ is rejected because an "arrow of time" has to be introduced. In a stochastic field of this nature we are involved only with the ensemble average. Thus Eq. (10.29) should be interpreted as follows: $\dot{a}_k(t)$ is the time differential of the *average* probability amplitude over an ensemble of atoms which have been prepared initially to have the same initial value $a_k(0)$. On the right-hand side we should substitute the ensemble average:

$$\langle\mathbf{E}^+_{ZP}(t)\cdot\mathbf{D}_{ki}\mathbf{D}_{ik}\cdot\mathbf{E}^-_{ZP}(t')\rangle_\varphi = \frac{D_{ki}D_{ik}\langle u_{ZP}(\tau)\rangle^+_\varphi}{3\varepsilon_0}$$

where $\langle u_{ZP}(\tau)\rangle_{\varphi}^{+}$ is the mutual coherence function defined in Eqs. (A2.4) and (A2.6). In this expression the factor 1/3 comes from the average value over all directions of $\cos^2 \beta$, where β is the angle between the polarization vector of $\mathbf{E}_{ZP}$ and the dipole moment vector of the atom $\mathbf{D}_{ki}$. The subscript on u_{ZP} reminds us that it is derived from the bilinear products of the zero-point field, and the superscript $^+$ that it is formed from the ordered product E^+E^- according to Eq. (A3.5). The averaging follows the path outlined in Appendix 3. The mutual coherence function of the zero-point field is, according to Eq. (A2.9),

$$\langle u_{ZP}(\tau)\rangle_{\varphi}^{+} = \frac{8\pi^2\varepsilon_0}{3}\int_0^{\infty} d\nu\, \nu^5 g^2(\nu)\, e^{-i2\pi\nu\tau}$$

For the case of the zero-point field, we have

$$g^2(\nu) = \frac{b^2}{\nu^2} = \frac{\mu_0}{4\pi}\frac{6h}{c\nu^2}$$

[refer to Eqs. (A2.13), (A3.14), and (10.18)]. Place these results in Eq. (10.29):

$$\dot{a}_k(t) = -\frac{\mu_0}{4\pi}\frac{64\pi^4 D_{ki}D_{ik}}{3ch}\int_0^t d\tau\, a_k(t-\tau)\int_0^{\infty} d\nu\, \nu^3\, e^{-i2\pi(\nu-\nu_0)\tau} \tag{10.30}$$

It can now be argued that $a_k(t-\tau)$ may be removed from the integrand and written as $a_k(t)$. This is because the rapid fluctuations of the exponential factor, when integrated over a large range of frequency, ensure that significant contributions come only from values of τ close to zero. Thus, the time derivative $\dot{a}_k(t)$ has only a short memory of the previous values of $a_k(t)$ between 0 and t; it remembers only the value of $a_k(t)$ immediately prior to t. The rate of change $a_k(t)$ depends rather on the fluctuations of the incoherent zero-point field whose coherence time is short—at the most important frequency, ν_0, the coherence time is of the order of $1/\nu_0$. Equation (10.30) can therefore be written as an ordinary differential equation:

$$\frac{\dot{a}_k(t)}{a_k(t)} = -\frac{\mu_0}{4\pi}\frac{64\pi^4 D_{ki}D_{ik}}{3ch}\int_0^{\infty} d\nu\, \nu^3 \int_0^t d\tau\, e^{-i2\pi(\nu-\nu_0)\tau} \tag{10.31}$$

Since we have argued that only small values of τ are significant, we can regard the upper limit of the time integral as infinite and write

$$\int_0^{\infty} d\tau\, e^{-i2\pi(\nu-\nu_0)\tau} = \tfrac{1}{2}\delta(\nu-\nu_0) + \frac{i}{2\pi}\mathcal{P}\left(\frac{1}{\nu-\nu_0}\right) \tag{10.32}$$

where $\mathcal{P}$ denotes that the principal value must be taken in any subsequent integration over frequency. Using this in Eq. (10.31), we derive the solution

$$\frac{\dot{a}_k(t)}{a_k(t)} = -\frac{\Gamma}{2} - i2\pi\Delta_L \tag{10.33}$$

where

$$\Gamma = \frac{\mu_0}{4\pi}\frac{64\pi^4\nu_0^3}{3ch}D_{ki}D_{ik} = A_{ki} \tag{10.34}$$

This result, deduced from semiclassical quantum theory, and using a zero-point vacuum field as a perturbation, is exactly the same as that achieved in Eq. (8.9) for the Einstein A coefficient using the pure classical phenomenological argument of Einstein and the correspondence principle to express the result in quantum mechanical terms. Following the processes used in Section 8.2 and above in Section 10.2, where the result is expanded to summation over all accessible lower states and substates, the standard form for the A coefficient, as given in Eq. (8.10) is achieved; and the decay constant is, as given in Eq. (8.12),

$$\gamma_k = \sum_{i<k}\sum_{M_i}\Gamma = \sum_{i<k} A_{ki} \tag{10.35}$$

Neglecting the last term in Eq. (10.33), we have the solution

$$a_k(t) = a_k(0)\, e^{-\gamma_k t/2}$$

and for the probability of the atom being found in state $|k\rangle$:

$$p_k(t) = p_k(0)\, e^{-\gamma_k t}$$

Equation (10.33) also indicates that a frequency shift of the transition occurs:

$$\Delta_L = \frac{\mu_0}{4\pi}\frac{16\pi^2}{3ch}D_{ki}D_{ik}\,\mathcal{P}\int_0^\infty d\nu\frac{\nu^3}{\nu - \nu_0} \tag{10.36}$$

Thus the same zero-point field required to secure Eq. (10.34) for the Einstein A coefficient, yields Eq. (10.36) for the unrenormalized first-order Lamb shift. It contains a highly divergent integral over frequency. The quantum electrodynamic treatment of this problem by H. Bethe,[3] obtains the renormalized solution with the factor ν^3 in the integrand replaced by ν_0^3 outside the integral; the divergence of this expression is much less severe (logarithmic) and can be handled in subsequent treatment. Some progress toward this result, although not a satisfying path mathematically, can be made by recognizing that a similar argument to that for placing $\tau \approx 0$ in Eq. (10.31) could be made for placing $\nu = \nu_0$; both of the quantities $(\nu - \nu_0)$ and τ occur as factors in the exponent in that equation, which contains an integral over ν as well as over τ. So that an argument for one of the quantities to be close to zero will apply also to the other. This could be of special importance in a small term like the Lamb shift of Eq. (10.36).

10.4. REMARKS ON SEMICLASSICAL TREATMENTS FOR SPONTANEOUS DECAY

In the preceding sections two apparently different methods have been used to account for spontaneous decay—the action of radiation reaction, and the action of the zero-point field. It would seem unreasonable that two entirely different causes should be responsible for the same effect. One suspects that the seat of radiation reaction must lie in the presence of the zero-point field, and vice versa; that they are, in fact, two manifestations of the same phenomenon. The matter will not be discussed further (the author does not know how to do it).

In every real atom the spontaneous decay by radiative transition of an excited level to one of lower energy, and eventually through subsequent decays to the ground level, is a fundamental property. There are, of course, cases where a level is a metastable state, the optical transition to the lower level being nonallowed by the radiative selection rules; for these cases the radiative decay constant, γ, is zero. In general, however, spontaneous decay is an intrinsic property of every excited state of the system and goes on independently of any other perturbation given to the system. One such perturbation, already discussed in Section 9.2, is the stimulation of transitions between two levels which is induced by electromagnetic radiation containing frequencies close to the natural frequency for transition between those two levels. This leads, for example, to Eq. (9.18) containing the Rabi frequency ν_R for the oscillation of probability between the two levels. In the deduction of that result, however, no allowance was made for the spontaneous decay of the upper level. The formula requires modification for the effect of spontaneous decay. This is not straightforward. Equations (9.9), the differential equations coupling the probability amplitudes a_k and a_i of the two-level system when decay is neglected, are relatively simple; in particular the phases of the two states are coherently related. On the other hand Eqs. (10.26) and (10.27), the differential equations for the probability amplitudes under spontaneous decay, are not so simple; they recognize the fact that the spontaneous decay of state $|k\rangle$ takes place to an infinite reservoir of states. These equations are irreversibility, whereas those for the Rabi transitions are, in essence, reversible.

It may be possible to study the complete phenomenon, in which case the ground state would have to be described as a superposition of the infinite reservoir of states into which the excited atom could decay; but it could prove useful to attempt to describe spontaneous decay in a more simple form, using the pretence that the ground level is describable by a single state function $|i\rangle$ with a probability amplitude a_i.

There is no difficulty in finding a differential equation for the probability amplitude of the excited state $|k\rangle$; it is basically Eq. (10.33). It can be rewritten as

$$\dot{a}_k = -\frac{\gamma}{2} a_k \tag{10.37}$$

where we have neglected the small frequency shift, and we have summed over all possible decays as in Eq. (10.35). (The subscript k has been dropped on γ.) This equation has the solution

$$a_k(t) = a_k(0)\, e^{-\gamma t/2} = A_k\, e^{-\gamma t/2} \tag{10.38}$$

where, for convenience, we have introduced $A_k = a_k(0)$, the amplitude of state $|k\rangle$ at $t = 0$. The probability of finding the atom in the state $|k\rangle$ at time t is given by

$$a_k(t)a_k^*(t) = A_k A_k^*\, e^{-\gamma t} \tag{10.39}$$

where $A_k A_k^*$ is the probability of finding the atom in the state $|k\rangle$ at $t = 0$.

To obtain an acceptable equation expressing the time development of the state $|i\rangle$ is not straightforward. We know, of course, from the requirements of normalization, that

$$a_i(t)a_i^*(t) = 1 - a_k(t)a_k^*(t) = 1 - A_k A_k^*\, e^{-\gamma t} \tag{10.40}$$

but this cannot readily be factored into two linear terms. Yet Eq. (10.27) implies that such a linear relationship exists between $\dot{a}_i(t)$ [or $\dot{b}_i(t)$] and $a_k(t)$. The proportionality is time dependent, of course, through the exponential oscillating factor and through the zero-point field $E_{ZP}(t)$, which contains an incoherent superposition of frequencies. It is postulated here that the differential equation can be expressed in the simplified form

$$\dot{a}_i(t) = +I\frac{\gamma}{2}\, a_k(t) \tag{10.41}$$

The $+\gamma/2$ factor is introduced so that the equation is dimensionally correct, and also in recognition of the fact that the rate of increase of a_i must be closely related to the rate of decrease of a_k. The dimensionless factor I is included to take account of all the incoherent factors in the phenomenon. The mathematical properties of I still have to be established; in some respects we are going to take the point of view that Eq. (10.41) is a recipe that will enable us to discuss the problem in simple terms without having to revert to the tedious difficulties of the earlier sections of this chapter. The justification for this lies solely in the success it has in producing sensible results. Firstly, I is to be regarded as a quantity that has been averaged over time and ensembles so that it is independent of t. Secondly, I must be "one-like" since, for the two-state case where $|k\rangle$ decays to $|i\rangle$, one would expect that the rate of growth of $|i\rangle$ equals the rate of decay of $|k\rangle$.

Formally Eq. (10.41) can be written, introducing the solution of Eq. (10.38), as

$$\dot{a}_i(t) = +I\frac{\gamma}{2} A_k e^{-\gamma t/2}$$

This can be integrated to give

$$a_i(t) = e^{i\alpha} - IA_k e^{-\gamma t/2} = e^{i\alpha} - Ia_k \tag{10.42}$$

where α is an unspecified phase, and the boundary condition $|a_i(t = \infty)| = 1$, representing the fact that, eventually, the atom will have decayed to the ground state, has been introduced [see comment after Eq. (11.23)]. Therefore we have

$$a_i(t)a_i^*(t)$$
$$= 1 - IA_k e^{-\gamma t/2} e^{-i\alpha} - I^*A_k^* e^{-\gamma t/2} e^{i\alpha} + II^*A_kA_k^* e^{-\gamma t} \tag{10.43}$$

This can be made to produce the acceptable result Eq. (10.40) if the following properties are defined for I:

> In any expression for probability, or other real observable quantity, linear terms in I and/or I^* (such as IA_k and $I^*A_k^*$, or their linear combination in the above) have the value zero. This reflects the fact that I and I^* incorporate the incoherencies of the physical situation, and averaging will give zero. (10.44)

> The product $II^* = -1$ (!). (10.45)

It can be seen that I and I^* are not usual mathematical quantities. It is as if I has the values $+1$ or -1 randomly so that when averaged over an ensemble its average value is zero. Thus $I^2 = 1$, $I^3 = I$, etc. But, whatever the value of I, the value of I^* is given by $I^* = -I$. Thus $II^* = -I^2 = -1$.

We shall use Eqs. (10.37) and (10.41) with due caution, to introduce decay into the two-state problem in the next chapter. Then, when a solution for a_k has been found, we shall use Eq. (10.42) to generate a solution for a_i. The justification for the use of the procedure lies in the success in generating acceptable results.

10.5. THE MATRIX ELEMENTS OF $\hat{H}_D$

The matrix elements of the Hamiltonian that expresses the spontaneous decay of the upper state of a two-state system can now be specified. Acting on the atom it will cause spontaneous decay of the upper state but will not affect the lower one. The following values:

$$\langle k|\hat{H}_D|k\rangle = \frac{-ih\Gamma}{4\pi}$$

$$\langle i|\hat{H}_D|k\rangle = I\frac{ih\Gamma}{4\pi}e^{i2\pi\nu_0 t} \tag{10.46}$$

$$\langle k|\hat{H}_D|i\rangle = \langle i|\hat{H}_D|i\rangle = 0$$

will ensure that, when the quantum mechanical equation of motion under $\hat{H}_A + \hat{H}_D$ is used with the expression Eq. (10.4) or similar for $|t\rangle$, Eqs. (10.37) and (10.41) will be obtained.

REFERENCES

1. D. T. PEGG, "Absorber Theory in Quantum Optics," *Phys. Scr.*, **T12**, 14–18 (1986).
2. T. H. BOYER, "Derivation of the Blackbody Radiation Spectrum Without Quantum Assumptions," *Phys. Rev.* **182**, 1374–1383 (1969).
3. H. A. BETHE, "The Electromagnetic Shift of Energy Levels," *Phys. Rev.* **72**, 339–341 (1947).

CHAPTER 11

THE GENERAL OPTICAL TRANSITION

11.1. THE TWO-LEVEL ATOM

In this chapter we discuss transitions, stimulated and spontaneous, in the two-level atom, described by an excited state $|k\rangle$ of energy $W_k = h\nu_k$, and a ground state $|i\rangle$ of energy $W_i = h\nu_i$. The transition frequency is $\nu_k - \nu_i = \nu_0$. Strictly speaking, of course, at least one of the two levels must comprise a number of sublevels or Zeeman levels. In the simplest of all transitions, for example, that between a $J = 0$ state and a $J = 1$ state, the latter has three substates with different projections of angular momentum $M = +1, 0, -1$. True two-state behavior can exist if the $J = 0$ state is the lower one or ground state, the $J = 1$ state is the upper one or excited state, and the radiation is polarized pure linear or pure circular. If the stimulating radiation is polarized pure linear, then we may regard the transition as being stimulated between the $J_i = 0$, $M_i = 0$ ground state and the $J_k = 1$, $M_k = 0$ excited state. Since spontaneous decay from the excited state can only return the atom to the unique ground state, the general state of the atom remains always a superposition of those two states. A similar argument can be made, for example when the radiation is pure circular; if it is designated σ^+, the ground state is coupled only to the $J_k = 1$, $M_k = 1$ excited state, and this can return only to $J_i = 0$, $M_i = 0$. Such a simple situation cannot occur for the case where the ground state is $J_i = 1$ and the excited state is $J_k = 0$. If, for example, one excites with linearly polarized light, an atom in the $J_i = 1$, $M_i = 0$ ground state is excited to the $J_k = 0$, $M_k = 0$ state; the latter may decay by spontaneous emission to any of the three ground states, $M_i = 0, \pm 1$, and the system has lost its two-state character. Indeed, unless there is some mechanism for returning the population to the $M_i = 0$ state, the atom will become optically pumped into the $J_i = 1$, $M_i = \pm 1$ states, because no transitions can be stimulated from these states. However, other cases of pure two-level behavior are possible. For example, if an atom whose ground state has angular momentum $J_i = J$ is stimulated to make transitions using σ^+-polarized radiation to an excited state with angular momentum $J_k = J + 1$, optical pumping will again occur. The combined effect of stimulation, which is always trying to transfer the atom to substates of higher component of angular momentum, and of spontaneous decay, is to pump the population by successive stimulations and decays, into the ground substate $J_i = J$, $M_i = +J$. From this, the incident radiation can stimulate transitions only to the excited substate $J_k = J + 1$, $M_k = J + 1$, and that state can return, by stimulated or spontaneous transition, only to $J_i = J$, $M_i = J$. The system, once pumped, can therefore be described as a superposition of these two states only.

We shall presume that the atomic states and the radiation polarization have been chosen such that pure two-state behavior is produced. The excited state $|k\rangle$ undergoes spontaneous decay to $|i\rangle$ with a rate constant γ. Transitions between the two levels are also stimulated by incident radiation; we consider initially the effect of a monochromatic field of frequency $\nu = \nu_0 + \delta$, the electric vector of which is described, as in Eq. (9.8), by

$$\mathbf{E}(t) = \frac{E_0}{\sqrt{2}}[\boldsymbol{\pi}^+ e^{-i2\pi\nu t} + \boldsymbol{\pi}^- e^{+i2\pi\nu t}] \tag{11.1}$$

The situation to be described is thus a superposition of the stimulated transitions given by the Rabi solutions of Chapter 9 and the spontaneous decay given in Chapter 10. We shall start, however, by reviewing the special case (which we shall refer to as the "Rabi" case) where spontaneous decay is neglected, $\gamma = 0$.

11.2. STIMULATED TRANSITIONS AT THE RABI FREQUENCY

Equation (9.9) gives the pair of coupled differential equations for the probability amplitudes a_k and a_i of the two states. They are rewritten here in slightly different notation:

$$\dot{a}_k = i\frac{\Omega}{2}e^{-i\Delta t}a_i \tag{11.2}$$

$$\dot{a}_i = i\frac{\Omega}{2}e^{+i\Delta t}a_k \tag{11.3}$$

For convenience the equations have been written in terms of angular frequencies:

$$\Omega = 2\pi\nu_R = \frac{2\sqrt{2}\,\pi}{h}E_0\boldsymbol{\pi}^+\cdot\mathbf{D}_{ki} = \frac{2\sqrt{2}\,\pi}{h}E_0\boldsymbol{\pi}^-\cdot\mathbf{D}_{ik}$$

$$= \text{the angular Rabi frequency} \tag{11.4}$$

$$\Delta = \omega - \omega_0 = 2\pi(\nu - \nu_0) = 2\pi\delta$$

= the difference between the incident field angular frequency and the transition angular frequency (11.5)

It should be recalled that these equations contain a small approximation—the rotating wave approximation—in that high-frequency terms in $\nu + \nu_0$ have been neglected.

It is worth pointing out that the two equations are consistent in the following sense. The rate of change of the probability $a_k a_k^*$ of finding the atom in state $|k\rangle$ can be calculated:

$$\frac{d}{dt}(a_k a_k^*) = \dot{a}_k a_k^* + a_k \dot{a}_k^* = i\frac{\Omega}{2}(e^{-i\Delta t} a_i a_k^* - e^{+i\Delta t} a_i^* a_k)$$

Similarly for the rate of change of the probability of finding the atom in state $|i\rangle$:

$$\frac{d}{dt}(a_i a_i^*) = \dot{a}_i a_i^* + a_i \dot{a}_i^* = i\frac{\Omega}{2}(e^{+i\Delta t} a_k a_i^* - e^{-i\Delta t} a_k^* a_i)$$

Therefore

$$\frac{d}{dt}(a_k a_k^* + a_i a_i^*) = 0$$

which is consistent with the normalization condition that, for a two-level system, the sum of the probabilities of the atom being found in the two levels must be constant (and equal to unity).

The second-order differential equation, obtained by eliminating a_i, can be written as ($D^2 \equiv d^2/dt^2$)

$$\left[D^2 - \left(\frac{i\Delta}{2}\right)^2\right] a_k\, e^{+i\Delta t/2} = +i\frac{\Omega}{2}\, \dot{a}_i\, e^{-i\Delta t/2} \tag{11.6}$$

On substitution of Eq. (11.3) for $\dot{a}_i$, one obtains [this is the same equation as Eq. (9.14)]

$$\left[D^2 - \left(\frac{i\Omega'}{2}\right)^2\right] a_k\, e^{+i\Delta t/2} = 0 \tag{11.7}$$

where the modified Rabi angular frequency Ω' is given by

$$\Omega' = [\Omega^2 + \Delta^2]^{1/2} \tag{11.8}$$

Equation (11.7) has the general solution

$$a_k = F e^{-i(\Delta-\Omega')t/2} + G e^{-i(\Delta+\Omega')t/2} \tag{11.9}$$

In Section 9.2, the initial condition $a_k(0) = 0$, $|a_i(0)| = 1$ was used. Here we shall introduce a more general initial condition so that we shall be able, at a later stage, to compare the result with the even more general one which includes exponential decay. Adopt the notation

$$a_k(0) = A_k$$

$$a_i(0) = A_i$$

These initial values of the probability amplitudes are related by the normalization condition:

$$A_k A_k^* + A_i A_i^* = 1 \tag{11.10}$$

There is no simple formula relating A_k and A_i but, by using the trick introduced in Section 10.4, we can write [cf. Eq. (10.42)]:

$$A_i = e^{i\alpha} - I A_k \tag{11.11}$$

and an equation that, according to the properties required of I, is consistent with it:

$$A_k = I\, e^{i\alpha} - I A_i = e^{i\beta} - I A_i$$

From this, using the properties of I, Eqs. (10.44) and (10.45), yield

$$\begin{aligned} A_i A_i^* &= 1 - (I A_k\, e^{-i\alpha} + I^* A_k^*\, e^{+i\alpha}) + I I^* A_k A_k^* \\ &= 1 - A_k A_k^* \end{aligned}$$

In this way Eq. (11.11) leads to the correct equation of normalization.

Either by placing the differential of Eq. (11.9) into Eq. (11.2), or by placing Eq. (11.9) into Eq. (11.3) and integrating, one obtains the corresponding solution for a_i:

$$a_i = \frac{\Omega' - \Delta}{\Omega} F e^{+i(\Delta+\Omega')t/2} - \frac{\Omega' + \Delta}{\Omega} G e^{+i(\Delta-\Omega')t/2} \tag{11.12}$$

or, what is equivalent in view of Eq. (11.8),

$$a_i = \frac{\Omega}{\Omega' + \Delta} F e^{+i(\Delta+\Omega')t/2} - \frac{\Omega}{\Omega' - \Delta} G e^{+i(\Delta-\Omega')t/2} \tag{11.12'}$$

Placing the initial conditions, one finds

$$\begin{aligned} F &= \frac{\Omega}{2\Omega'} A_i + \frac{\Omega' + \Delta}{2\Omega'} A_k \\ G &= \frac{-\Omega}{2\Omega'} A_i + \frac{\Omega' - \Delta}{2\Omega'} A_k \end{aligned} \tag{11.13}$$

The solution, Eq. (11.9), for the probability amplitude, a_k, can now be written as follows:

$$\begin{aligned} a_k = &\left(\frac{\Omega}{2\Omega'} A_i + \frac{\Omega' + \Delta}{2\Omega'} A_k\right) e^{-i(\Delta-\Omega')t/2} \\ &+ \left(\frac{-\Omega}{2\Omega'} A_i + \frac{\Omega' - \Delta}{2\Omega'} A_k\right) e^{-i(\Delta+\Omega')t/2} \end{aligned} \tag{11.14}$$

The probability of finding the atom in the state $|k\rangle$ at time t is

$$a_k a_k^* = A_k A_k^* + \frac{\Omega^2}{\Omega'^2}(1 - 2A_k A_k^*) \sin^2\left(\frac{\Omega'}{2} t\right) \tag{11.15}$$

In reaching this result certain products have been placed equal to zero: $A_k A_i^*$, $A_k^* A_i$. This is because, even initially, the phases of the two states $|k\rangle$ and $|i\rangle$ will be random. Over an ensemble, and even for a single atom where the phase relation is unknown, an average over all possible phases will make these products zero. There may be occasions, where the initial state of the atoms are precisely prepared, when the more complete expression than Eq. (11.15) could be required.

We note that, when $A_k = 0$ and $|A_i| = 1$, Eq. (11.14) reduces to

$$a_k = \frac{i\Omega}{\Omega'} e^{-i\Delta t/2} \sin\left(\frac{\Omega'}{2} t\right)$$

which is the same as the expression already deduced in Chapter 9, [Eq. (9.17)]. Equation (11.15) reduces to

$$a_k a_k^* = \frac{\Omega^2}{\Omega'^2} \sin^2\left(\frac{\Omega'}{2} t\right)$$

which is the same as Eq. (9.18).

It is unnecessary to resort to any artifice by the use of the quantity I in order to obtain an expression for a_i. From either Eq. (11.12) or (11.12′) we obtain

$$a_i = \left(\frac{\Omega' - \Delta}{2\Omega'} A_i + \frac{\Omega}{2\Omega'} A_k\right) e^{i(\Delta+\Omega')t/2} + \left(\frac{\Omega' + \Delta}{2\Omega'} A_i - \frac{\Omega}{2\Omega'} A_k\right) e^{i(\Delta-\Omega')t/2} \tag{11.16}$$

$$a_i a_i^* = 1 - A_k A_k^* - \frac{\Omega^2}{\Omega'^2}(1 - 2A_k A_k^*) \sin^2\left(\frac{\Omega'}{2} t\right) \tag{11.17}$$

in agreement with Eq. (11.15) and the requirement of normalization.

11.3. THE TWO-LEVEL ATOM WITH DECAY

The coupled differential equations for the motion of the system without decay, $\gamma = 0$, are given by Eqs. (11.2) and (11.3); they have been demonstrated to be compatible with the requirement of normalization of probability. They lead to the solutions Eq. (11.14) and Eq. (11.16) for a_k and a_i, respectively.

The differential equations for pure decay have been discussed in Section 10.4; they are Eqs. (10.37) and (10.41):

$$\dot{a}_k = -\frac{\gamma}{2} a_k \tag{11.18}$$

$$\dot{a}_i = +I\frac{\gamma}{2} a_k \tag{11.19}$$

The first of these is not coupled, expressing the fact that state $|k\rangle$ decays of its own accord. The second expresses the fact that, as state $|k\rangle$ decays, the probability amplitude a_i grows at a rate that depends on the amount of $|k\rangle$ present. However, there is no relation between the phases of the two states in this incoherent decay process, as compared with the stimulated transition where the states are linked in phase by a coherent driving field. The role of the factor I in allowing for this state of affairs is discussed in Section 10.4. It should be noted that, since I can equally be $+1$ or -1, Eq. (11.19) would be equally valid with a negative sign on the right-hand side, or indeed with I^* on the right-hand side. It must be noted that the two equations for $\dot{a}_k$ and $\dot{a}_i$ are compatible with the requirements of normalization *only if the solution for exponential decay is used*; we could not expect that it would be possible to incorporate Eq. (11.19) into a more general equation and still lead to a sensible solution for a_i. We shall, however, meld Eqs. (11.2) and (11.18) to give

$$\dot{a}_k = i\frac{\Omega}{2}e^{-i\Delta t}a_i - \frac{\gamma}{2}a_k \tag{11.20}$$

It is not, at this point, clear how to write a differential equation for a_i. We shall adopt, however, for the second-order differential equation for a_k, a generalization of Eq. (11.7):

$$\left[D^2 - \left(\pm\frac{i\Omega'}{2} + \frac{\gamma}{4}\right)^2\right]a_k\, e^{i\Delta t/2}\, e^{\gamma t/4} = 0 \tag{11.21}$$

Certainly, when $\gamma = 0$, this reduces to Eq. (11.7). And when $\Omega'^2 = 0$ (i.e., $\Omega = 0$ and $\Delta = 0$), the equation leads to exponential decay. One can work back from these equations to find a differential equation for a_i. That equation is

$$\dot{a}_i = \frac{i\Omega}{2}e^{+i\Delta t}a_k + \left(\frac{\pm\Omega' + \Delta}{\Omega}\right)\frac{\gamma}{2}e^{i\Delta t}a_k \tag{11.22}$$

The first term on the right-hand side is just that expected for Rabi-type transitions—see Eq. (11.3).

The second term is, presumably, an appropriate generalization to express the effect of spontaneous decay. The factor $e^{i\Delta t}$ is to be expected; it maintains the proper phase relation between a_i and a_k as in the first term of the differential equation. This phase relation is implied also in Eq. (11.20). Its inclusion into Eq. (10.42) would lead that equation to be written as

$$a_i = e^{i\alpha} - I\, e^{i\Delta t}a_k \tag{11.23}$$

The inclusion of the extra factor makes no difference to the normalization relationship $a_i a_i^* = 1 - a_k a_k^*$. Comparison of Eq. (11.22) with Eq. (11.19) suggests that we must identify $[(\pm\Omega' + \Delta)/\Omega]$ with I. We shall adopt the following identification for later use:

$$\begin{aligned} &\text{When the positive value of } \Omega' \text{ is used:} \quad && +\Omega' + \Delta = I\Omega \\ &\text{When the negative value of } \Omega' \text{ is used:} \quad && -\Omega' + \Delta = I^*\Omega \end{aligned} \tag{11.24}$$

Note that the product of these two equations, $-\Omega'^2 + \Delta^2 = II^*\Omega^2 = -\Omega^2$, is consistent with the relationship between Ω' and Ω, i.e., $\Omega'^2 = \Omega^2 + \Delta^2$. However, this artifice will not be needed in the main arguments to follow, but only in one of the tests on the validity of the final result.

We accept therefore the differential equation (11.21) as describing the motion of the system. Depending on which sign is used, it yields two solutions:

$$\begin{aligned} a_{k+} &= e^{-i\Delta t/2}[F_+ \, e^{i\Omega' t/2} + G_+ \, e^{-i\Omega' t/2} \, e^{-\gamma t/2}] \\ a_{k-} &= e^{-i\Delta t/2}[F_- \, e^{-i\Omega' t/2} + G_- \, e^{i\Omega' t/2} \, e^{-\gamma t/2}] \end{aligned} \tag{11.25}$$

Both equations lead eventually to the same final result; we shall use the first only and shall drop the subscript plus. Expressions for the coefficients F and G are obtained by applying the initial conditions, $a_k(t = 0) = A_k$, $a_i(t = 0) = A_i$:

$$\begin{aligned} F &= \frac{i\Omega A_i + i(\Omega' + \Delta)A_k}{i2\Omega' + \gamma} \\ G &= \frac{-i\Omega A_i + [i(\Omega' - \Delta) + \gamma]A_k}{i2\Omega' + \gamma} \end{aligned} \tag{11.26}$$

The complete expression for a_k is

$$\begin{aligned} a_k = e^{-i\Delta t/2}\left\{\frac{i\Omega A_i + i(\Omega' + \Delta)A_k}{i2\Omega' + \gamma} \, e^{+i\Omega' t/2}\right. \\ \left. + \frac{-i\Omega A_i + [i(\Omega' - \Delta) + \gamma]A_k}{i2\Omega' + \gamma} \, e^{-i\Omega' t/2} \, e^{-\gamma t/2}\right\} \end{aligned} \tag{11.27}$$

and a_i can be found from Eq. (11.23), $a_i = e^{i\alpha} - I \, e^{i\Delta t} a_k$. The probability of finding the atom in the upper state $|k\rangle$ is

$$a_k a_k^* = FF^* + (FG^* \, e^{i\Omega' t} + F^* G \, e^{-i\Omega' t}) \, e^{-\gamma t/2} + GG^* \, e^{-\gamma t} \tag{11.28}$$

and $a_i a_i^*$ can be found from $a_i a_i^* = 1 - a_k a_k^*$. Expressions for the coefficients are

$$FF^* = \frac{\Omega^2 A_i A_i^* + (\Omega' + \Delta)^2 A_k A_k^*}{4\Omega'^2 + \gamma^2}$$

$$= \frac{\Omega^2 + 2\Delta(\Omega' + \Delta) A_k A_k^*}{4\Omega'^2 + \gamma^2}$$

$$FG^* = \frac{-\Omega^2 A_i A_i^* + [(\Omega'^2 - \Delta^2) + i(\Omega' + \Delta)\gamma] A_k A_k^*}{4\Omega'^2 + \gamma^2}$$

$$= \frac{-\Omega^2 + [2\Omega^2 + i(\Omega' + \Delta)\gamma] A_k A_k^*}{4\Omega'^2 + \gamma^2} \tag{11.29}$$

$$F^*G = (FG^*)^* = \frac{-\Omega^2 + [2\Omega^2 - i(\Omega' + \Delta)\gamma] A_k A_k^*}{4\Omega'^2 + \gamma^2}$$

$$GG^* = \frac{\Omega^2 A_i A_i^* + [(\Omega' - \Delta)^2 + \gamma^2] A_k A_k^*}{4\Omega'^2 + \gamma^2}$$

$$= \frac{\Omega^2 + [2\Delta(-\Omega' + \Delta) + \gamma^2] A_k A_k^*}{4\Omega'^2 + \gamma^2}$$

In forming these products it has been recognized that there is no coherent phase relationship between the initial amplitudes A_i and A_k. The cross products have therefore been placed equal to zero.

We shall carry out three tests on Eq. (11.28) to check that it gives expected results.

Test 1. At $t = 0$, Eq. (11.28) reduces to

$$a_k a_k^* = FF^* + FG^* + F^*G + GG^* = A_k A_k^*$$

in agreement with the definition of A_k.

Test 2. When $\gamma = 0$, Eq. (11.28) reduces to

$$a_k a_k^* = FF^* + GG^* + (FG^* e^{i\Omega' t} + F^*G\, e^{-i\Omega' t})$$

$$= A_k A_k^* + \frac{\Omega^2}{\Omega'^2}(1 - 2A_k A_k^*) \sin^2\left(\frac{\Omega' t}{2}\right)$$

in agreement with the Rabi result deduced in Eq. (11.15).

Test 3. When $\Omega = 0$ one would hope to produce the equation for natural, spontaneous decay of the excited state. However, the differential equation, Eq. (11.21), is not constituted to produce that result without recourse to the artifice expressed in Eq. (11.24). If Eq. (11.24) is used (i.e.,

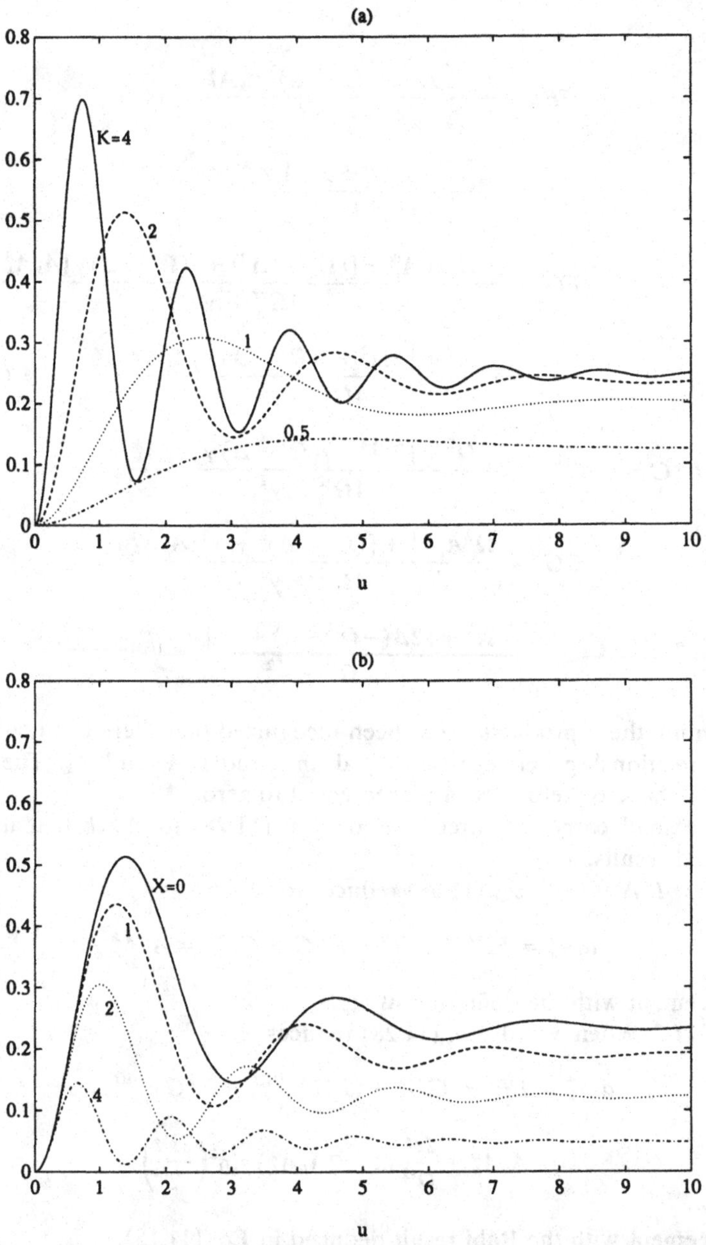

FIGURE 12. The probability of excitation to an excited state by monochromatic light as a function of time ($u = \gamma t$). (a) When the driving field is resonant ($X = 0$). The graphs are drawn for $K = 2\pi\nu_R/\gamma = 4, 2, 1, 0.5$. (b) When $K = 2$. The graphs are drawn for $X = 2\pi\delta/\gamma = 2\pi(\nu - \nu_0)/\gamma = 0, 1, 2, 4$.

$\Omega' + \Delta = I\Omega$) in Eq. (11.26) and Ω is then placed equal to zero:

$$F = 0, \qquad G = A_k$$

Then $a_k = A_k\, e^{-\gamma t/2}$, and $a_k a_k^* = A_k A_k^*\, e^{-\gamma t}$.

The success in deducing previously established results gives confidence that Eq. (11.28) represents a valid general solution (it does not prove it, however).

The normal experimental circumstance would be that the atom rests, before stimulation, in the ground state: $a_k(t = 0) = A_k = 0$. The probability $p(t) = a_k a_k^*$ that the atom is to be found in state $|k\rangle$ at time t becomes

$$\begin{aligned} p(t) = a_k a_k^* &= \frac{\Omega^2}{4\Omega'^2 + \gamma^2}[1 - (e^{+i\Omega' t} + e^{-i\Omega' t})\, e^{-\gamma t/2} + e^{-\gamma t}] \\ &= \frac{4\pi^2 \nu_R^2}{16\pi^2 \nu_R'^2 + \gamma^2}(1 - 2\, e^{-\gamma t/2} \cos 2\pi\nu_R' t + e^{-\gamma t}) \end{aligned} \tag{11.30}$$

As $t \to \infty$, a steady state is reached for which the probability of the atom being found in the excited state $|k\rangle$ is $\Omega^2/(4\Omega'^2 + \gamma^2)$. It must be remembered that this result is for monochromatic radiation of angular frequency $\omega_L = \omega_0 + \Delta$, and of intensity [see Eq. (9.12)] $I = h^2\Omega^2/8\pi^2\mu_0 c D^2$, (where $D^2 = \mathbf{D}_{ki} \cdot \mathbf{D}_{ik}$). In Fig. 12a, for the case where the driving field is resonant (i.e., $\Delta = 0$), $p(t)$ is plotted against time for various values of $\Omega = 2\pi\nu_R$ (i.e., for various strengths of the driving field). Figure 12b shows the effect of nonresonant driving field.

11.4. THE STEADY-STATE SOLUTION

Equation (11.30) shows that, at time long compared with the mean radiative lifetime $1/\gamma$, the transients and the Rabi oscillations have died. The oscillations will still be occurring in individual atoms, but remember that we are dealing with an ensemble. The oscillations are apparent immediately after switching on the monochromatic perturbation when all the atoms are responding in phase; but as individual atoms decay and are restimulated the phases of the Rabi oscillations become random. The steady-state probability is (we call this the Rabi probability indicated by the superscript R):

$$p_k^R = a_k a_k^* = \frac{\Omega^2}{4\Omega'^2 + \gamma^2} = \frac{\Omega^2}{4\Omega^2 + 4\Delta^2 + \gamma^2} \tag{11.31}$$

It may be asked how this result relates to the population of states as deduced from the Einstein coefficients. From Eq. (8.3), the probability of finding the atom in level k (we call this the Einstein probability indicated by the superscript E) should be

$$p_k^E = \frac{N_k}{N_i + N_k} = \frac{B_{ik}\rho(\nu_0)}{B_{ik}\rho(\nu_0) + B_{ki}\rho(\nu_0) + A_{ki}} \tag{11.32}$$

The circumstances under which these results have been derived are, at the moment, quite different. The excited state probability derived from the Rabi treatment, $a_k a_k^*$, is for excitation of a *strictly* two-state atom by a *monochromatic* beam of radiation of frequency $\nu_L = \nu_0 + \Delta/2\pi$. The probability derived from the Einstein treatment, p_k^E, is for excitation of a two-level system (but with substates accounted for in the summations in the expressions for A and B, Eqs. (8.10), (8.19), and (8.20)), by radiation which has a *broad-band spectrum*, at least compared to the resonant width of the atomic transition. However, one should be able to integrate Eq. (11.31) over all directions of radiation in space, over all polarization directions, and over a wide spectral range. This should then achieve the result of Eq. (11.32) appropriate to excitation by isotropic, unpolarized, broad-band radiation.

We shall discuss this under the condition of weak excitation, $\Omega \ll \gamma$, $B\rho \ll A$. We would certainly expect to achieve some sort of agreement under this regime. Of course, it may be felt that the Einstein treatment is not strictly limited strictly to low-intensity excitation; however, it is certainly the case that the Einstein treatment does not recognize the oscillating probability that emerges from the Rabi treatment. For this reason we limit discussion to the low-intensity, weak excitation case. In this circumstance, the probabilities are

$$p_k^R = a_k a_k^* = \frac{\pi^2 \nu_R^2}{4\pi^2(\nu_L - \nu_0)^2 + \gamma^2/4} \tag{11.33}$$

$$p_k^E = \frac{B_{ik}\rho(\nu_0)}{\gamma} \tag{11.34}$$

We now prepare to integrate Eq. (11.33) over frequency, space, and polarization. We recall Eq. (9.12) for the Rabi frequency:

$$\nu_R^2 = \frac{2\mu_0 c}{h^2}(\boldsymbol{\pi}_\lambda^+ \cdot \mathbf{D}_{ki})(\boldsymbol{\pi}_\lambda^- \cdot \mathbf{D}_{ik})\frac{s(\nu)}{4\pi}\, d\nu \sin\theta\, d\theta\, d\phi \tag{11.35}$$

in which I, the intensity of the monochromatic beam, has been replaced by $[s(\nu)/4\pi]\, d\nu \sin\theta\, d\theta\, d\phi$; $s(\nu)$ is the *spectral intensity* defined in Eq. (A1.18) and will itself be replaced by $c\rho(\nu)$, where $\rho(\nu)$ is the *spectral energy density* defined in Eq. (A1.16); $d\nu$ is the differential spectral window of the differential beam; $\sin\theta\, d\theta\, d\phi$ is the differential solid angle of the beam, which has been normalized by the factor of 4π in the denominator. The label λ on the polarization vectors prepares for summation over states of polarization. Equation (11.35) can be written as

$$\nu_R^2 = \frac{\mu_0 c^2 D^2}{2\pi h^2} \cos^2 \alpha_\lambda\, \rho(\nu)\, d\nu \sin\theta\, d\theta\, d\phi \tag{11.36}$$

where α_λ is the angle between the polarization vector $\boldsymbol{\pi}_\lambda^+$ of the differential beam and the vector character of transition electric-dipole matrix element $\mathbf{D}_{ki} = \langle k|\hat{\mathbf{D}}|i\rangle$ in the coordinate axes of quantization being used to describe the atomic states. Equation (11.33) can now be integrated to give the probability of finding the atom in the state $|k\rangle$ under isotropic, unpolarized, broad-band excitation:

$$p_k^R = \frac{\pi\mu_0 c^2 D^2}{2h^2} \sum_\lambda \int_0^\infty d\nu \int_0^\pi d\theta \int_0^{2\pi} d\phi$$

$$\times \frac{1}{4\pi^2(\nu-\nu_0)^2 + \gamma^2/4} \cos^2 \alpha_\lambda\, \rho(\nu) \sin\theta \tag{11.37}$$

The summation is over two orthogonal states of polarization for each differential beam, each with its own value of α_λ.

It is now necessary to specify α_λ in terms of the angles θ and ϕ. Figure 3 in Section 3.2, which shows the relationships between various axes, angles, and unit vectors, will aid discussion. A beam of radiation propagating along the θ, ϕ (or $\mathbf{r}^0$) direction is polarized in the $\boldsymbol{\theta}^0$–$\boldsymbol{\phi}^0$ plane. Any two orthogonal vectors in that plane can be used to describe the polarization of the incoherent light; $\boldsymbol{\theta}^0$ and $\boldsymbol{\phi}^0$ will be suitable if a π-transition ($\Delta M = 0$) is being discussed; $2^{-1/2}(\boldsymbol{\theta}^0 + i\boldsymbol{\phi}^0)$ and $2^{-1/2}(\boldsymbol{\theta}^0 - i\boldsymbol{\phi}^0)$ will be suitable if a σ-transition ($\Delta M = \pm 1$) is being discussed—see discussion of polarization in Section 3.2.

Of course, since we are preparing to integrate over all angles, polarizations, and frequencies, we can no longer have a situation of two-state behavior. Therefore, we eventually have to sum over all transitions between the lower and upper levels. To concentrate our attention, and with the knowledge that the sum rules will lead to the same final result when the sum over all transitions is taken, we shall discuss, in particular, a simple transition—a lower level of one state, $J_i = 0$, $M_i = 0$, and an excited level of three states, $J_k = 1$, $M_k = +1, 0, -1$. Summation can eventually be made over the three transitions involved, which, in this case, all have the same probability. For this transition, $\mathbf{D}_{ki} = \langle k|\hat{\mathbf{D}}|i\rangle = \langle J_k M_k|\hat{\mathbf{D}}|J_i M_i\rangle$, and the vector characters are given by (see Eqs. 7.23 and 7.24)

$$\langle 1, +1|\hat{\mathbf{D}}|0, 0\rangle = \frac{1}{\sqrt{2}}(\mathbf{x}^0 - i\mathbf{y}^0)D$$

$$\langle 1, 0|\hat{\mathbf{D}}|0, 0\rangle = \mathbf{z}^0 D$$

$$\langle 1, -1|\hat{\mathbf{D}}|0, 0\rangle = \frac{1}{\sqrt{2}}(\mathbf{x}^0 + i\mathbf{y}^0)D$$

where D is the reduced matrix element.

Consider, firstly, the π-transition having the vector character $\mathbf{z}^0$. For this case it is simplest to use for the two orthogonal polarizations:

$$\lambda = 1: \quad \boldsymbol{\pi}_1^+ = \boldsymbol{\theta}^0$$

$$\lambda = 2; \quad \boldsymbol{\pi}_2^+ = \boldsymbol{\phi}^0$$

$$\cos\alpha_1 = \frac{\boldsymbol{\pi}_1^+ \cdot \mathbf{D}_{ki}}{D} = \boldsymbol{\theta}^0 \cdot \mathbf{z}^0 = \cos\theta: \quad \cos^2\alpha_1 = \cos^2\theta$$

$$\cos\alpha_2 = \frac{\boldsymbol{\pi}_2^+ \cdot \mathbf{D}_{ki}}{D} = \boldsymbol{\phi}^0 \cdot \mathbf{z}^0 = 0: \quad \cos^2\alpha_2 = 0$$

The integration in Eq. (11.37) over angles, and the summation over polarization thus reduces to

$$\int_0^{\pi} d\theta \int_0^{2\pi} d\phi \cos^2\theta \sin\theta = \frac{4\pi}{3}$$

To illustrate that the same result is reached for a σ transition, consider the σ^+ transition for which the vector character is $2^{-1/2}(\mathbf{x}^0 - i\mathbf{y}^0)$. The appropriate choice of basis vectors for polarization is

$$\lambda = 1: \quad \boldsymbol{\pi}_1^+ = \frac{1}{\sqrt{2}}(\boldsymbol{\theta}^0 + i\boldsymbol{\phi}^0)$$

$$\lambda = 2: \quad \boldsymbol{\pi}_2^+ = \frac{1}{\sqrt{2}}(\boldsymbol{\theta}^0 - i\boldsymbol{\phi}^0)$$

Then, we have

$$\cos\alpha_1 = \tfrac{1}{2}(1 + \cos\theta)\, e^{-i\phi} \quad \text{and} \quad \cos^2\alpha_1 = \tfrac{1}{4}(1 + \cos\theta)^2$$

and similarly,

$$\cos\alpha_2 = \tfrac{1}{2}(1 - \cos\theta)\, e^{-i\phi} \quad \text{and} \quad \cos^2\alpha_2 = \tfrac{1}{4}(1 - \cos\theta)^2$$

The summation and integration gives

$$\int_0^{\pi} d\theta \int_0^{2\pi} d\phi\, \tfrac{1}{2}(1+\cos^2\theta)\sin\theta = \frac{4\pi}{3}$$

It is worth noting, at this point, the consequence of the above values for $\cos\alpha_1$ and $\cos\alpha_2$. If the radiation is along the z axis, $\theta = 0$, $\cos^2\alpha_1 = 1$ and $\cos^2\alpha_2 = 0$. Thus one choice of circular polarization stimulates fully the $\Delta M = 1$ transition; the other has no effect.

The integration of angles and the summation of polarizations in Eq. (11.37) have now been carried out; the result is

$$p_k^R = \frac{2\pi^2\mu_0 c^2 D^2 \rho(\nu_0)}{3h^2} \int_0^{\infty} d\nu \frac{1}{4\pi^2(\nu-\nu_0)^2+\gamma^2/4} \tag{11.38}$$

where the energy density has been taken outside the integration because it is presumed to be reasonably constant over the width of the Lorentzian resonance. Carrying out the integration yields

$$p_k^R = \frac{2\pi^2\mu_0 c^2 D^2 \rho(\nu_0)}{3h^2\gamma} \tag{11.39}$$

This is the population probability of each of the substates of the excited level; the total population probability of the level is three times the above. The result has been established by starting with the Rabi expression for the oscillating transitions between levels, taking into account the effect of spontaneous decay, finding the steady state, and, finally, integrating over isotropic, unpolarized, broad-band radiation. How does the result agree with that of the Einstein treatment? From Eq. (11.34), using Eq. (8.20) for B_{ik}, neglecting the summation so as to find the probability of finding the atom in *one* of the excited M states:

$$p_k^E = \frac{2\pi^2\mu_0 c^2 D^2 \rho(\nu_0)}{3h^2\gamma} \tag{11.40}$$

in exact agreement.

This analysis has provided a further powerful test for the validity of the results in Section 11.3. We accept Eqs. (11.28) and (11.29) as correct expressions for the probability for finding the atom in the upper state, with Eq. (11.30) as the special case when the atom starts initially in the ground state.

11.5. RADIATION FROM A DRIVEN ATOM

For an atom that is driven by a monochromatic beam of radiation (laser beam) of frequency ν_L so that the probability of it being found in the excited state is $a_k a_k^*$, one would expect that the reradiated, scattered, or fluorescence field would be

$$E(t) \propto a_k a_k^* \, e^{-i2\pi\nu_L t} + \text{c.c.} \tag{11.41}$$

Since the atoms are being excited to a state of energy $h\nu_L$ above the ground level, with strong or resonant excitation if $\nu_L \approx \nu_0 = \nu_{ki}$, they will reradiate at that frequency. Since the probability of the atom being in that state varies with time according to Eq. (11.28) or, more specifically, Eq. (11.30), and those equations contain an oscillating term, the radiated field will be *amplitude modulated* and therefore will contain side-bands at $\nu_L \pm \nu'_R$. Equation (11.41) can be further justified as follows. The expression for the general state of the atom is

$$|t\rangle = a_k \, e^{-i2\pi\nu_0 t}|k\rangle + a_i|i\rangle$$

The expectation value of the dipole moment of the atom is therefore

$$\langle t|\hat{\mathbf{D}}|t\rangle = a_k a_i^* \langle i|\hat{\mathbf{D}}|k\rangle \, e^{-i2\pi\nu_0 t} + \text{c.c.}$$

Suppressing the geometrical factors and the vector character that denotes the polarization, the radiation field of the induced dipole is [compare the classical expression Eq. (3.35)]

$$E(t) \propto a_k a_i^* D \, e^{-i2\pi\nu_0 t} + \text{c.c.}$$

Using Eq. (11.23), we obtain

$$E(t) \propto e^{-i\alpha} a_k D \, e^{-i2\pi\nu_0 t} - I^* a_k a_k^* D \, e^{-i2\pi\nu_L t} + \text{c.c.} \tag{11.42}$$

If we neglect the complications of the factor I^* and place it equal to -1, the second term is in agreement with Eq. (11.41); we shall discuss the contribution of the first term later. Using Eq. (11.30) and Eq. (11.41), we obtain

$$E(t) \propto A(1 - 2\, e^{-\gamma t/2} \cos 2\pi\nu'_R t + e^{-\gamma t}) \cos 2\pi\nu_L t \tag{11.43}$$

where

$$A = \frac{4\pi^2\nu_R^2}{16\pi^2\nu_R^2 + 16\pi^2\delta^2 + \gamma^2} \qquad \text{for } t > 0$$

$$= 0 \qquad \text{for } t < 0$$

and

$$\nu'_R = (\nu_R^2 + \delta^2)^{1/2}$$

The Fourier transform, Eq. (3.17), of this is

$$\tilde{E}(\nu) \propto A \int_0^\infty (1 - 2\,e^{-\gamma t/2} \cos 2\pi\nu'_R t + e^{-\gamma t}) \cos 2\pi\nu_L t\, e^{i2\pi\nu t}\, dt \quad (11.44)$$

Expressing this in analytic form and retaining only the resonant terms in $\nu - \nu_L$ (i.e., rejecting the high-frequency terms in $\nu + \nu_L$ which will integrate to give small contributions only), we obtain

$$\tilde{E}(\nu) \propto \frac{A}{2}\left[\delta(\nu - \nu_L) + \frac{1}{-i2\pi(\nu - \nu_L) + \gamma} - \frac{1}{-i2\pi(\nu - \nu_L - \nu'_R) + \gamma/2} - \frac{1}{-i2\pi(\nu - \nu_L + \nu'_R) + \gamma/2}\right] \quad (11.45)$$

This represents a δ-function spike at $\nu = \nu_L$ and three Lorentzian resonances (see Appendix 5) at $\nu = \nu_L,\ \nu_L \pm \nu'_R$. The spectral intensity [see Eq. (4.23)] of this scattered field is determined from

$$\frac{\tilde{E}(\nu)\tilde{E}^*(\nu)}{\mu_0 cT} \propto \frac{A^2}{4\mu_0 cT\gamma}\left\{\left(2\gamma T + 2 - \frac{2\gamma^2}{4\pi^2\nu_R^2 + \gamma^2/4}\right)\delta(\nu - \nu_L) + \left[\frac{\gamma}{4\pi^2(\nu - \nu_L)^2 + \gamma^2} + \frac{\gamma}{4\pi^2(\nu - \nu_L - \nu'_R)^2 + \gamma^2/4} + \frac{\gamma}{4\pi^2(\nu - \nu_L + \nu'_R)^2 + \gamma^2/4}\right]\right\} \quad (11.46)$$

The term in large parentheses comes from the products of $\delta(\nu - \nu_L)$ with itself and the other three terms. In forming these products it is anticipated that the result, $s(\nu)$, will eventually be integrated over ν and, consequently, $\delta(\nu - \nu_L)$ operates on the other factors according to Eq. (A5.9). In the case of $[\delta(\nu - \nu_L)]^2$, this becomes $\delta(0)\delta(\nu - \nu_L) = 2T\delta(\nu - \nu_L)$ in accordance with the properties of the peaked functions discussed in Appendix 5 before the limit of large T is taken.

The term in large square brackets consists only of the products of the last three terms in Eq. (11.45) with their own complex conjugate. The other cross-product terms are ignored on the grounds that when one factor is resonant the other is not and therefore is small. This is really true only if $\nu'_R \gg \gamma$, but it will serve as a useful approximation because, in any experiment designed to observe these separate terms, one must excite the atoms strongly to achieve that condition. Under this condition of strong excitation the third term in the large parentheses is small and can also be neglected.

The expression shows the presence of side-bands at frequencies $\nu_L \pm \nu'_R$ on either side of the central scattered frequency of the laser ν_L. The effect is called the *dynamic or ac Stark effect* because it is caused by the oscillating electric field in the driving radiation.

According to Eqs. (4.23), (A1.16), and (A1.18) the spectral intensity of the reradiated or scattered field is given by this expression in the limit as $T \to \infty$. However, in this limit all but the first term has vanished. In practice, as has been pointed out previously (in Section 5.3 for example), if information about a frequency f is to be retained in the signal, the integration time constant T must be kept short enough that $fT < 1$. If information about the side-bands, displaced by $\pm\nu'_R$ from the laser frequency, is to be preserved, then T must be rather less than $1/\nu'_R$. Furthermore, if the side-bands are to be resolved clearly from resonances that have a natural width γ (let alone any Doppler width to be commented on below) then ν'_R must be several times γ; or $1/\nu'_R$ must itself be rather less than $1/\gamma$. Therefore, in a practical experiment to study the dynamic Stark effect, the smoothing time constant in the detector circuit would be chosen at least a factor of 10 smaller than $1/\gamma$; therefore $\gamma T < 0.1$ and is probably considerably smaller still. Hence we write $T \to$ "∞," implying that T is to be taken "great enough to average over optical frequencies and statistical fluctuations, but not so great as to destroy the side-bands." Under these circumstances, it is the first term only in the curly brackets of Eq. (11.46) that disappears. The spectral intensity is

$$s(\nu) \propto \lim_{T \to \text{``}\infty\text{''}} \frac{A^2}{\mu_0 c \gamma T} \left[2\delta(\nu - \nu_L) + \frac{\gamma}{4\pi^2(\nu - \nu_L)^2 + \gamma^2} + \frac{\gamma}{4\pi^2(\nu - \nu_L + \nu'_R)^2 + \gamma^2/4} + \frac{\gamma}{4\pi^2(\nu - \nu_L - \nu'_R)^2 + \gamma^2/4} \cdots \right] \tag{11.47}$$

where the ellipsis indicates other terms that are being omitted and neglected. The four surviving terms in the square brackets are the principal resonances—a "spike" at $\nu = \nu_L$, and three Lorentzian resonances centered on $\nu = \nu_L$, $\nu_L \pm \nu'_R$.

The two side-band resonances have heights of four times the central Lorentzian and half the width. Their integrated areas are twice the center one, although, with the delta function included, the three resonances have integrated areas of 1:2.5:1, respectively. This does not accord with the analysis of, e.g., Mollow,[1] or Carmichael and Walls.[2] These authors have analyzed a more general problem where both of the states, upper and lower, are subject to decay, where the "width" of the states have contributions from other processes (collisions), and where quantum fluctuations are taken into account. For the case of pure radiative decay from an upper state to a ground state and for strong excitation, i.e., where the conditions are similar to those used in the treatment above, their side-bands are of one-third the height of the center, and these three resonances have integrated area 1:2:1. Their results are more in agreement with experiments (e.g., Schuda *et al.*[3] Grove *et al.*[4]). The situation vis à vis comparison between theory and experiment is not clear, however. Theory has basically assumed that the field stimulating the atoms is monochromatic. There will, inevitably, be some spectral width to this field, and this will be grossly affected by the thermal motion of the atoms, providing a Doppler broadening. The Lorentzian resonances will be modified to a Voigt-type profile (i.e., a convolution of a Lorentzian and a Gaussian) with reduced height and broader profile. The central "spike" will also be broadened to a Doppler profile, which will superimpose on the central Lorentzian. Then the predicted appearance would become more like the observations.

Furthermore, there is a contribution from the first term in Eq. (11.42). Because of the factor $e^{-i\alpha}$ the contributions from this are incoherent with respect to those from the second term. Nonetheless the term [using Eq. (11.27) with $A_k = 0$ and $A_i = 1$]

$$E_\alpha(t) \propto a_k\, e^{-i2\pi\nu_0 t} = \frac{i2\pi\nu_R}{i4\pi\nu'_R + \gamma} \times [e^{-i2\pi(\nu_0+\delta/2-\nu'_R/2)t} + e^{-i2\pi(\nu_0+\delta/2+\nu'_R/2)t}\, e^{-\gamma t/2}]$$

will generate, when integration over the frequency distribution is carried out, some resonant-type contributions rather closer to the central frequency. With incomplete resolution these will add to the size of the central resonance. Also, because of the explicit dependence on $\delta = \nu_L - \nu_0$, these terms could contribute to nonsymmetric resonance if the laser is not tuned accurately to the atomic frequency.

There is a further contribution to scattering at the laser frequency or within a Doppler shift thereof. All electrons in the atom, and in any atom that may be bathed in the incident radiation, will scatter coherently by the Rayleigh scattering process. For all these reasons it will prove difficult to predict accurately the relative size of the central resonance and the side-bands.

The total physical effect of the scattering by an atom of radiation that is very intense and nearly resonant with an absorption transition is therefore very complicated. The principal feature is the appearance in the scattered radiation of side-bands at $\nu_L - \nu'_R$ and $\nu_L + \nu'_R$. The semiclassical analysis presented here certainly leads to a prediction of these features that have been experimentally observed. It is, however, rather simplified.

REFERENCES

1. B. R. Mollow, "Power Spectrum of Light Scattered by Two-Level Systems," *Phys. Rev.* **188**, 1969–1975 (1969).
2. H. J. Carmichael and D. F. Walls, "A Quantum-Mechanical Master Equation Treatment of the Dynamical Stark Effect," *J. Phys. B: Atom. Mol. Phys.* **9**, 1199–1219 (1967).
3. F. Schuda, C. R. Stroud Jr., and M. Hercher, "Observation of the Resonant Stark Effect at Optical Frequencies", *J. Phys. B: Atom. Mol. Phys.* 7, L198–202 (1974).
4. R. E. Grove, F. Y. Wu, and S. Ezekiel, "Measurement of the Spectrum of Resonance Fluorescence from a Two-Level Atom in an Intense Monochromatic Field," *Phys. Rev. A* **15**, 227–233 (1977).

CHAPTER 12

THE PHOTOELECTRIC EFFECT

12.1. A GENERAL DISCUSSION OF THE PHOTOELECTRIC EFFECT

We are aware of light (the term "light" is used here to include a wide range of electromagnetic radiations)—or perhaps we postulate with conviction the existence of a physical entity that we chose to call light—because we observe the effects when it interacts with, or is absorbed by, electrons, atoms, and matter. The sun radiates light; we feel its warmth on our skin, which we attribute to its infrared content; we suffer the effect of its destructive influence when we get sunburnt, and we attribute this to its ultraviolet content; but most obviously, we detect the radiations within a certain frequency window by the receptors in our eyes, which are tuned to recognize a certain spectrum of frequencies in the light. And, apart from our own personal experiences through our senses, we can build detectors that can provide information on the spectral distribution of intensity and frequency of radiation.

As has been described in earlier chapters, the effect of the absorption of light or radiation is to raise the energy of the atoms in the absorber, and this effect can be used as the basis of a *physical* detector of the radiation. Since light is an electromagnetic radiation is interacts with charged particles. The most effective interaction is with electrons because they are plentiful and of low mass. As has already been discussed in Chapter 6, on the Compton effect, the interaction transfers energy and momentum to the electron. However, the Compton effect refers specifically to the interaction with *free* electrons and *they* are not abundant. To attempt to use this effect as a light detection process would not produce significant results. Interaction of light with electrons in atoms (or molecules or matter) is much more promising. It has already been established that when light of frequency ν interacts with an atom it raises the energy state of an electron by $h\nu$. What physical change takes place depends on how this energy compares with the energy structure of the absorber.

If the absorber is a nonconducting solid or liquid, the electrons are in states that are bound to atoms or molecules of the material. If $h\nu$ is small an electron may simply be raised to another bound state, a process that can be *resonant* (and therefore highly likely to occur) if $h\nu$ corresponds closely to an energy difference between allowed states. The atoms may then lose this energy by reradiation, either at the same frequency (*resonance fluorescence*) or at reduced frequencies if transitions through intermediate states are involved (*cascade fluorescence*).

If the absorber is a semiconductor the electron may be excited to one of the levels in the conducting continuum. The material then becomes momentarily conducting and, with appropriate circuitry, can deliver a pulse of current—the physical detection of radiation is achieved by this *photoconductive* effect.

Should the value of $h\nu$ be large enough, the electron may be released from the material into a free state. Let W_I be the minimum energy needed to raise the electron from its normal state in the material, either a bound state in a nonconductor or a conduction level in a conductor, to a state in which it is free to move away from the material; W_I is often referred to as the *workfunction* or *ionization energy* of the material. To create a free electron, $h\nu$ must be greater than W_I; any excess of $h\nu$ over W_I will become kinetic energy of the free electron:

$$K = h\nu - W_I \tag{12.1}$$

This phenomenon—the absorption of energy $h\nu$ from the light raising the energy of an electron from a bound state into a free state—is called the *photoelectric effect.* Since the presence of a single free electron can readily be detected by making it initiate an avalanche of ionizations in an appropriate gas over which there is an electric field, thereby producing a pulse of current, a single-quantum absorptive transition can be detected.

Although, from the point of view of making practical devices, the absorber will usually be condensed matter, the same process occurs in single atoms and molecules. We shall discuss the photoelectric process in an atom.

Figure 13 represents energy levels of a typical atom. At the bottom are the filled shells, with the outer valence electron in its ground level of state vector $|i\rangle$. Of course, excitation of any of the bound electrons can occur, in which case W_I is the ionization energy from that bound level. We concentrate on excitation of the outer electron, whose initial state we now take to be the zero of energy. Above this are the excited bound levels of the atom with state vectors $|j\rangle$ having energies $W_j = h\nu_j$ above the ground. If the atom is bathed in light of (near) monochromatic frequency ν_L, (the subscript L implying "light" or "laser"), this will give strong excitation to any state $|k\rangle$ that has an energy $W_k = h\nu_k$ close to the quantum energy

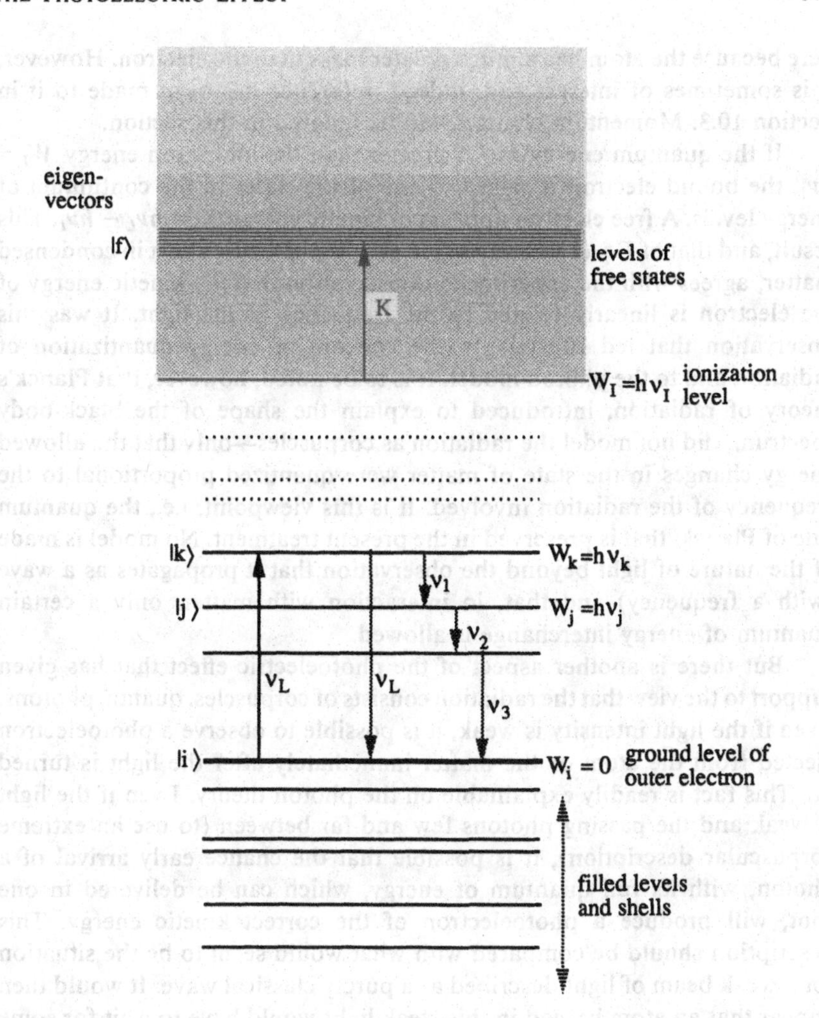

FIGURE 13. The energy levels of a typical atom—filled levels, the ground level, bound excited levels, the ionization level, continuum of free levels.

change $h\nu_L$, provided that the selection rules allow the transition. As we have seen previously, the energy of excitation is $h\nu_L$ (not $h\nu_k$); but the probability of excitation is most strong when these are nearly equal. After excitation the atom may make the reverse transition to the ground state by stimulated emission and/or spontaneous decay. Alternatively, the decay may be through a cascade of spontaneous transitions through lower levels shown in the diagram as $\nu_1 \rightarrow \nu_2 \rightarrow \nu_3$. In all these absorption and emission processes there will be transfer of momentum as well as of energy. Whereas this is of considerable importance in the Compton effect, it is unimportant here because the atom has a much greater mass than the electron. However, it is sometimes of interest and, indeed, reference has been made to it in Section 10.3. Momentum changes will be ignored in this section.

If the quantum energy $h\nu_L$ is greater than the ionization energy $W_I = h\nu_I$, the bound electron is raised to one of the states in the continuum of energy levels. A free electron appears of kinetic energy $K = h\nu_L - h\nu_I$. This result, and that of Eq. (12.1) relating to the photoelectric effect in condensed matter, agrees with the experimental observation that the kinetic energy of the electron is linearly related to the frequency of the light. It was this observation that led Einstein to the concept of energy quantization of radiation and to the photon model. It is to be noted, however, that Planck's theory of radiation, introduced to explain the shape of the black-body spectrum, did not model the radiation as corpuscles—only that the allowed energy changes in the state of matter were quantized proportional to the frequency of the radiation involved. It is this viewpoint, i.e., the quantum rule of Planck, that is preserved in the present treatment. No model is made of the nature of light beyond the observation that it propagates as a wave (with a frequency) and that, in interaction with matter, only a certain quantum of energy interchange is allowed.

But there is another aspect of the photoelectric effect that has given support to the view that the radiation consists of corpuscles, quanta, photons. Even if the light intensity is weak, it is possible to observe a photoelectron ejected from the atom or the matter immediately after the light is turned on. This fact is readily explainable on the photon theory. Even if the light is weak and the passing photons few and far between (to use an extreme corpuscular description), it is possible that the chance early arrival of a photon, with its full quantum of energy, which can be delivered in one shot, will produce a photoelectron of the correct kinetic energy. This description should be compared with what would seem to be the situation for a weak beam of light described as a purely classical wave. It would then appear that an atom bathed in this weak light would have to wait for some time before it had "soaked up" enough of the passing energy to emit an electron into a free state; in which case there would always be a measurable time lag between the onset of the light and the appearance of the first photoelectron. Observation, however, shows that there is some possibility of observing the first photoelectron "with no detectable time lag" or, at least, in times shorter than predicted by classical argument. This could be expected from the photon picture, where the first corpuscle might very well "pass through the effective target area" of the atom almost immediately after switch-on of the light even if one has to wait some time for the next one to arrive. At first sight, the phenomenon appears to favor the corpuscular model. However, we are not here dealing with the purely classical wave picture. Also, we should avoid the type of description, typified by the words in quotation marks above in this paragraph, that is often used in textbooks and tends to mislead students toward favoring the photon picture (see, for example, the excellent textbook by Halliday and Resnick[1]).

The same thing, i.e., the possibility of the immediate excitation of an electron, is predicted by the *semi*classical model. This fact can be illustrated by reference to Fig. 10 in Section 9.2. The graph shows that, even for weak intensity and short time, there is some finite probability that the absorbing atom will be found in the excited level—it does not have to "wait" until it has "soaked up" enough energy from the radiation field. The probability may be low, but the fact that it is not zero means that it will occasionally occur. And that is exactly the same thing that is predicted by the photon picture: if the radiation is weak, then the photons pass infrequently, and one would expect to wait some time on average to intercept a photon, but occasionally an earlier arrival will occur. Both models predict the same thing. The language one uses depends on one's preference of model—either that the energy is localized in the field in quanta of $h\nu$, with the attendant problem of the nature of this localization, or that there are quantum rules for the interaction, with the attendant problem that, at any instant, energy may not have been conserved.

We are going to extend the theory of Chapter 9 to the photoelectric effect and deduce an expression for the cross section.

12.2. THE PHOTOELECTRIC DIFFERENTIAL CROSS SECTION

We shall first deduce an expression for the differential cross section starting from Eq. (9.18) for the probability of finding the atom in an excited state $|k\rangle$ at time t:

$$p_k(t) = \frac{\nu_R^2}{\nu_R^2 + \delta^2} \sin^2[\pi(\nu_R^2 + \delta^2)^{1/2} t] \tag{12.2}$$

where $\delta = \nu_L - \nu_k$ is the off-resonance frequency and ν_R is the Rabi frequency given by [see Eq. (9.12)]

$$\nu_R^2 = \frac{2E_0^2}{h^2}(\boldsymbol{\pi}^+ \cdot \mathbf{D}_{ki})(\boldsymbol{\pi}^- \cdot \mathbf{D}_{ik}) = \frac{2\mu_0 ce^2 I_i}{h^2}(\boldsymbol{\pi}^+ \cdot \mathbf{r}_{ki})(\boldsymbol{\pi}^- \cdot \mathbf{r}_{ik}) \tag{12.3}$$

$I_i = E_0^2/\mu_0 c$ is the intensity of an incident monochromatic beam. Equation (12.2) assumes that spontaneous decay from state $|k\rangle$ can be neglected compared with stimulated transition back to the initial state $|i\rangle$. However, we are going to make the presumption that the time t is short compared with the Rabi period ($t \ll 1/2\pi\nu_R$) so that only the initial rise of the probability is significant.

When the light frequency ν_L is greater than the ionization threshold frequency $\nu_I = W_I/h$, the excited state reached is not an isolated bound state but a state of the continuum. All states of the continuum, provided they are accessible from $|i\rangle$ by electric dipole transition of the appropriate polarization, can be excited. Note that, *for any state excited*, the electron has a kinetic energy given by Eq. (12.1), but only those states whose eigenfrequency ν_k is close to ν_L will be strongly excited. Since the accessible states form a continuum we must integrate over all frequencies ν_k from the lower bound of the excited continuum ν_I to infinity. The probability per unit solid angle is

$$\frac{dp(t)}{d\Omega} = \int_{\nu_I}^{\infty} p_k(t) \frac{dn(\nu_k)}{d\Omega}\, d\nu_k \tag{12.4}$$

where $n(\nu_k)$ is the number of free states of frequency ν_k—i.e., of kinetic energy $K = h(\nu_k - \nu_I)$—per unit frequency interval and in solid angle $d\Omega$; thus the equation is expressed in terms of the probability per unit solid angle and, in general, will be a function of angle or direction in space of photoelectric emission. Replacing ν_k by $\delta = \nu_k - \nu_L$, we obtain

$$\frac{dp(t)}{d\Omega} = \int_{-\infty}^{\infty} p_k(t) \frac{dn(\nu_k)}{d\Omega}\, d\delta \tag{12.5}$$

The lower limit is strictly $\delta = -(\nu_L - \nu_I)$. This has been replaced by infinity because $p_k(t)$ is resonant at $\delta = 0$ with a width of ν_R; see Eq. (12.2). Provided the light frequency ν_L is greater than the ionization frequency by several times the Rabi frequency, which presumption we now make, the substitution is well justified. The state density $dn(\nu_k)/d\Omega$ can be regarded as reasonably constant through the narrow resonance; it can be replaced by its value at the light frequency, $dn(\nu_L)/d\Omega$, and brought outside the integral. Furthermore, although ν_R is a function of ν_k (and hence of δ) through its dependence on the matrix elements D_{ki} and D_{ik}, it is also a slowly varying function of δ and can be removed from the integral, its value, as given in Eq. (12.3), being replaced by

$$\nu_R^2 = \frac{2E_0^2}{h^2}(\boldsymbol{\pi}^+ \cdot \mathbf{D}_{Li})(\boldsymbol{\pi}^- \cdot \mathbf{D}_{iL}) = \frac{2\mu_0 c e^2 I_i}{h^2}(\boldsymbol{\pi}^+ \cdot \mathbf{r}_{Li})(\boldsymbol{\pi}^- \cdot \mathbf{r}_{iL}) \tag{12.6}$$

The matrix element $\mathbf{r}_{iL}$ is now to be evaluated between the *initial ground state* of the atom $|i\rangle$ and the *free state* $|L\rangle$. Equation (12.5) can now be rewritten. Using Eq. (12.2), the probability that the atom will be found to have undergone photoelectric emission into unit solid angle at time t is

$$\frac{dp(t)}{d\Omega} = \frac{dn(\nu_L)}{d\Omega} \nu_R^2 \int_{-\infty}^{\infty} \frac{1}{\nu_R^2 + \delta^2} \sin^2[\pi(\nu_R^2 + \delta^2)^{1/2} t]\, d\delta$$

$$= \frac{dn(\nu_L)}{d\Omega} \frac{\pi^2 \nu_R^2}{\gamma} \left[\frac{4}{\pi} \int_0^{\infty} \frac{\sin^2[(K^2 + x^2)^{1/2} u/2]}{K^2 + x^2}\, dx \right] \tag{12.7}$$

in which we have written $K = 2\pi\nu_R/\gamma$, $x = 2\pi\delta/\gamma$, $u = \gamma t$ in the integral, where γ is some rate constant used to make the quantities in the integration dimensionless. This has been done in order to make the equation closely compatible with a later equation, Eq. (12.22).

The rate at which the free electron is produced (per atom) is

$$\frac{d}{dt}\left[\frac{dp(t)}{d\Omega}\right] = \frac{dn(\nu_L)}{d\Omega} \pi\nu_R^2 \int_{-\infty}^{\infty} \frac{\sin[2\pi t(\nu_R^2 + \delta^2)^{1/2}]}{(\nu_R^2 + \delta^2)^{1/2}}\, d\delta$$

$$= \frac{dn(\nu_R)}{d\Omega} \pi^2 \nu_R^2 \left[\frac{2}{\pi} \int_0^{\infty} \frac{\sin[(K^2 + x^2)^{1/2} u]}{(K^2 + x^2)^{1/2}}\, dx \right]$$

$$= \frac{dn(\nu_L)}{d\Omega} \pi^2 \nu_R^2 J_0(2\pi\nu_R t) \tag{12.8}$$

The two quantities defined in square brackets in Eqs. (12.7) and (12.8) are illustrated in the graphs of Fig. 14 where we have placed $K = 1$, i.e., the chosen rate constant is $\gamma = 2\pi\nu_R$ and the time axis in units of $Ku = 2\pi\nu_R t$.

For short times compared with the Rabi period, as mentioned above, we have

$$\frac{d}{dt}\left[\frac{dp(t)}{d\Omega}\right]_{t=0} = \frac{dn(\nu_L)}{d\Omega} \pi^2 \nu_R^2 \tag{12.9}$$

Using Eq. (12.6) for ν_R^2, the rate at which free electrons are produced per unit solid angle per atom is

$$\frac{d}{dt}\left[\frac{dp(t)}{d\Omega}\right]_{t=0} = \frac{4\pi^2\alpha}{h} I_i \frac{dn(\nu_L)}{d\Omega} \langle L|\boldsymbol{\pi}^+ \cdot \hat{\mathbf{r}}|i\rangle\langle i|\boldsymbol{\pi}^- \cdot \hat{\mathbf{r}}|L\rangle \tag{12.10}$$

where the fine structure constant $\alpha = \mu_0 c e^2/2h = 1/137.036\ldots$ has been introduced. The probability for photoelectric emission is proportional to the intensity of the incident light.

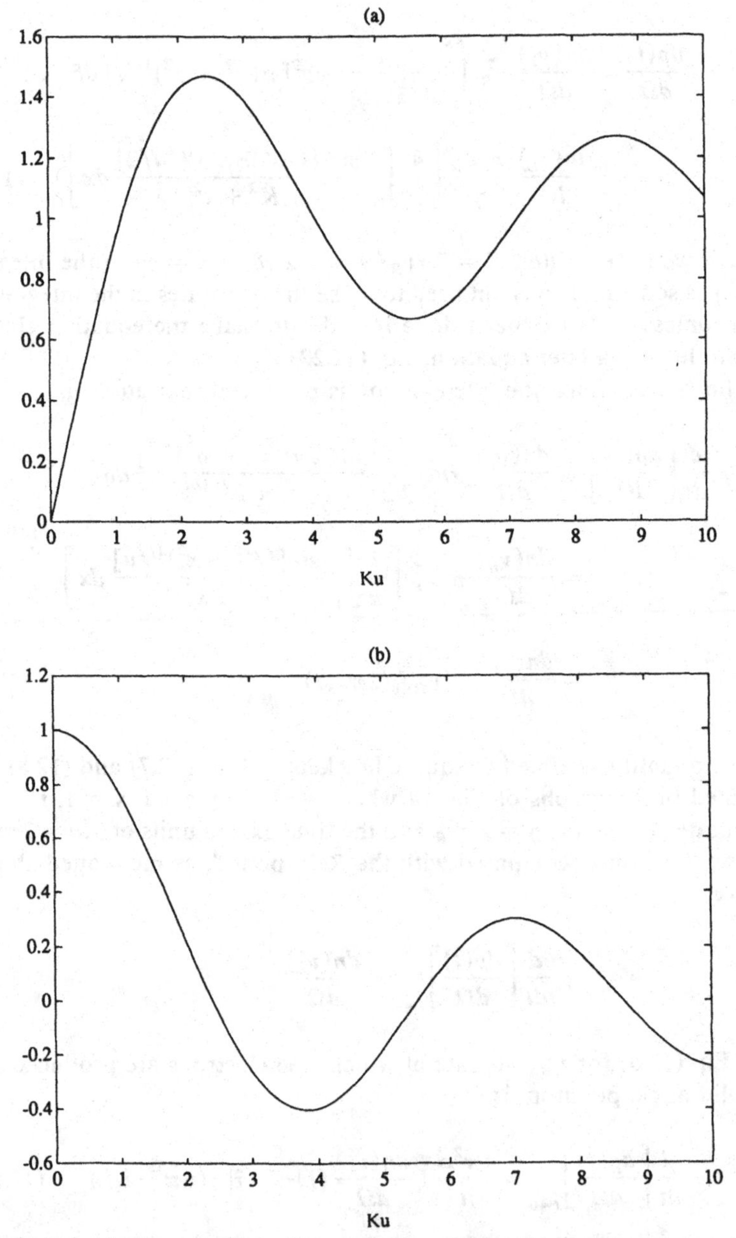

FIGURE 14. (a) The probability of photoelectric emission. (b) The rate of photoelectric emission. Each graph is drawn as a function of $Ku = 2\pi\nu_R t$, where the Rabi frequency, ν_R, is proportional to the square root of the intensity of the light.

The differential cross section for photoelectric emission is

$$\frac{d\sigma}{d\Omega} = \frac{\text{rate of emission of free electron energy per unit solid angle}}{\text{intensity of the incident light}}$$

$$= \frac{\text{rate of emission of free electrons per unit solid angle} \times \text{free electron energy}}{\text{intensity of the incident light}}$$

$$= \frac{\dfrac{d}{dt}\left[\dfrac{dp(t)}{d\Omega}\right]_{t=0} \times h(\nu_L - \nu_I)}{I_i}$$

$$= 4\pi^2\alpha(\nu_L - \nu_I)\frac{dn(\nu_L)}{d\Omega}\langle L|\boldsymbol{\pi}^+ \cdot \hat{\mathbf{r}}|i\rangle\langle i|\boldsymbol{\pi}^- \cdot \hat{\mathbf{r}}|L\rangle \tag{12.11}$$

The result of Eq. (12.11) will be further developed below when it is applied to the photoelectric effect in the hydrogen atom. But before this, it is instructive to deduce a result for the photoelectric cross-section using a rather different model and to study the more general case. To reiterate, the above model uses light strong enough in intensity so that the spontaneous decay process, or any cascade of events that leads eventually to the recapture of the freed electron into the initial ground state, can be neglected. However, the result is appropriate for short times (short compared with the Rabi period) so that no substantial buildup of free electron probability has occurred. Of course, once the electron is freed and thereby participating in the production of a photoelectric signal, the theory no longer applies. The ionized atom will, at some later time, capture another electron from somewhere and the process of photoelectric excitation will begin again. The above theory is therefore applicable to the onset of photoelectric excitation following the moment when the interaction of light with the atom in the initial state $|i\rangle$ commences.

12.3. THE PHOTOELECTRIC DIFFERENTIAL CROSS SECTION—AN ALTERNATIVE VIEW

In this model we make the assumption that the intensity of the exciting light is weak. Under this condition, stimulated emission process is much less probable than the spontaneous capture process, which is described by a decay constant Γ_k; $p_k\Gamma_k$ is the rate at which the atom, with probability p_k of being in the excited state $|k\rangle$, returns to ground.

Consider an atom in its ground state $|i\rangle$ and subjected to electromagnetic radiation with an E field given by

$$\mathbf{E}(t) = \frac{E_0}{\sqrt{2}}[\boldsymbol{\pi}^+ e^{-i2\pi\nu_L t} + \boldsymbol{\pi}^- e^{i2\pi\nu_L t}] \tag{12.12}$$

which is of rms amplitude E_0 and which is linearly polarized if $\boldsymbol{\pi}^+ = \boldsymbol{\pi}^- = \boldsymbol{\pi}^0$, and is circularly polarized in the **ij** plane if $\boldsymbol{\pi}^{\pm} = (\mathbf{i} \pm i\mathbf{j})/\sqrt{2}$. We consider the case where $\nu_L > \nu_I = W_I/h$, where W_I is the ionization energy from state $|i\rangle$. Absorption of energy $h\nu_L$ from the radiation raises the atom to one of the free states $|k\rangle$ in the continuum with an energy

$$W_L = h\nu_L = K + h\nu_I$$

where K is the kinetic energy of the free electron. We note, again, that the kinetic energy of the electron is well defined according to the quantum rule for interaction, but that the particular state of the continuum into which the electron is excited is not defined. Many states of excitation are possible, the most probable being those whose eigenenergies $h\nu_k$ above ground are close to $h\nu_L$; see Fig. 15.

Since the interaction is weak, the interaction Hamiltonian

$$\hat{H}_{\text{int}} = -\mathbf{E}(t) \cdot \hat{\mathbf{D}} = e\mathbf{E}(t) \cdot \hat{\mathbf{r}}$$

is a weak perturbation to the Hamiltonian $\hat{H}_A$, which describes the internal energy of the atom, and to the Hamiltonian $\hat{H}_D$, which describes the decay of the electrons from the continuum state.

The general state of the atom can be expressed as a superposition of eigenstates of the atomic Hamiltonian with time-dependent amplitudes:

$$|t\rangle = \sum_k a_k(t)\, e^{-i2\pi\nu_k t}|k\rangle + a_i|i\rangle$$

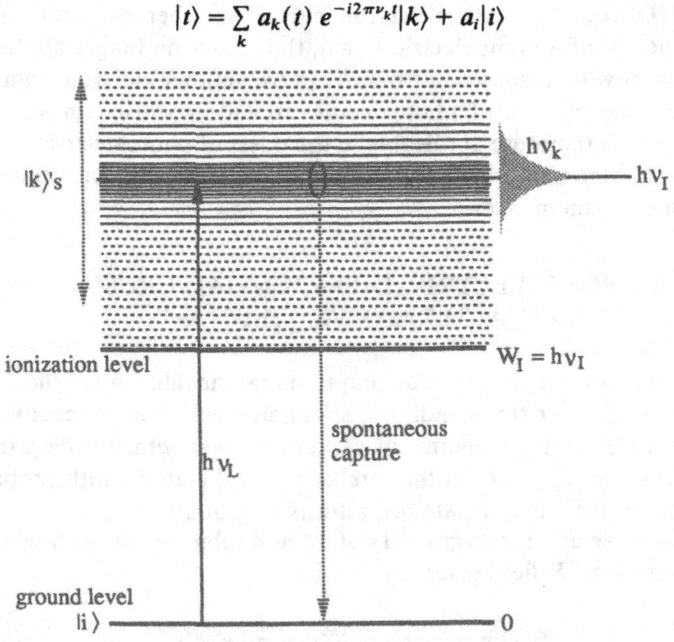

FIGURE 15. Excitation of an atom into the continuum.

The fact that the excited state lies in the continuum is here expressed by the summation; eventually we shall replace this by an integration.

Solve the equation of motion:

$$\frac{ih}{2\pi}\frac{\partial}{\partial t}|t\rangle = \hat{H}|t\rangle \qquad \text{with } \hat{H} = \hat{H}_A + \hat{H}_D + \hat{H}_{\text{int}}$$

The matrix elements of $\hat{H}_D$ are given in Eq. (10.46). Treatment by the usual methods yields the differential equations for the amplitudes:

$$\dot{a}_k + \frac{\Gamma_k}{2}a_k = \frac{2\pi}{ih}e^{i2\pi\nu_k t}\,e\mathbf{E}(t)\cdot\langle k|\hat{\mathbf{r}}|i\rangle a_i$$
$$\dot{a}_i - I\frac{\Gamma_k}{2}a_k = \frac{2\pi}{ih}\sum_k e^{-i2\pi\nu_k t}\,e\mathbf{E}(t)\cdot\langle i|\hat{\mathbf{r}}|k\rangle a_k \qquad (12.13)$$

In the second equation I is the mathematical quantity introduced in Section 10.4; in fact we shall not use it here. Substitute Eq. (12.12) for the electric field and retain only the slowly varying term in the low frequency $\nu_L - \nu_k$ (the "antiresonant" term with the frequency $\nu_L + \nu_k$ will become negligible on integration). Write $a_k = \alpha_k e^{-\Gamma_k t/2}$; then $\dot{a}_k + (\Gamma_k/2)a_k = \dot{\alpha}_k e^{-\Gamma_k t/2}$. The first equation now reads

$$\dot{\alpha}_k = \frac{2\pi}{ih}e^{\Gamma_k t/2}\,e^{i2\pi(\nu_k-\nu_L)t}\frac{eE_0}{\sqrt{2}}\langle k|\boldsymbol{\pi}^+\cdot\hat{\mathbf{r}}|i\rangle a_i \qquad (12.14)$$

Use the weak perturbation assumption, i.e., that, when the steady state is reached, all $a_k \ll 1$, and $a_i \approx 1$. The equation can then be integrated from $t = -\infty$ to t to yield

$$\alpha_k(t) = \frac{2\pi}{h}\frac{eE_0}{\sqrt{2}}\langle k|\boldsymbol{\pi}^+\cdot\hat{\mathbf{r}}|i\rangle\frac{e^{-i2\pi(\nu_L-\nu_k)t}\,e^{\Gamma_k t/2}}{2\pi(\nu_L-\nu_k)+i\Gamma_k/2} \qquad (12.15)$$

The expression for $|t\rangle$ becomes

$$|t\rangle = \sum_k \frac{\sqrt{2}\,\pi eE_0\langle k|\boldsymbol{\pi}^+\cdot\hat{\mathbf{r}}|i\rangle}{h[2\pi(\nu_L-\nu_k)+i\Gamma_k/2]}e^{-i2\pi\nu_L t}|k\rangle + |i\rangle \qquad (12.16)$$

The probability that the atom is found in the free state $|k\rangle$ is

$$p(t) = \sum_k \frac{2\pi^2e^2E_0^2}{h^2}\frac{1}{4\pi^2(\nu_L-\nu_k)^2+\Gamma_k^2/4}\langle k|\boldsymbol{\pi}^+\cdot\hat{\mathbf{r}}|i\rangle\langle i|\boldsymbol{\pi}^-\cdot\hat{\mathbf{r}}|k\rangle \quad (12.17)$$

In the steady state, the probability of photoelectric emission per unit time equals the probability of electron capture per unit time by the ionized atom. Therefore the probability of photoelectric emission per unit time (the rate) is

$$R(t) = \sum_k \Gamma_k p_k(t)$$
$$= \frac{2\pi^2 e^2 E_0^2}{h^2} \sum_k \frac{\Gamma_k \langle k|\boldsymbol{\pi}^+ \cdot \hat{\mathbf{r}}|i\rangle\langle i|\boldsymbol{\pi}^- \cdot \hat{\mathbf{r}}|k\rangle}{4\pi^2(\nu_L - \nu_k)^2 + \Gamma_k^2/4} \tag{12.18}$$

This summation over all states of the continuum can be replaced by an integration if we enter the density of states:

$\frac{dn(\nu_k)}{d\Omega}$ = the number of states of electron kinetic energy $h(\nu_k - \nu_I)$ per unit frequency interval per unit solid angle

The rate of emission of free electrons per unit solid angle is then

$$\frac{dR(t)}{d\Omega} = \frac{2\pi^2 e^2 E_0^2}{h^2} \int_{\nu_I}^{\infty} d\nu_k \frac{dn(\nu_k)}{d\Omega} \frac{\Gamma_k \langle k|\boldsymbol{\pi}^+ \cdot \hat{\mathbf{r}}|i\rangle\langle i|\boldsymbol{\pi}^- \cdot \hat{\mathbf{r}}|k\rangle}{4\pi^2(\nu_L - \nu_k)^2 + \Gamma_k^2/4} \tag{12.19}$$

Using, as we did in Section 12.2, the fact that the state density and the matrix elements are reasonably constant over the region of the resonance, they can be taken out of the integration and replaced by their values at $\nu_k = \nu_L$. After performing the integration we obtain

$$\frac{dR(t)}{d\Omega} = \frac{dn(\nu_L)}{d\Omega} \frac{2\pi^2 e^2 E_0^2}{h^2} \langle L|\boldsymbol{\pi}^+ \cdot \hat{\mathbf{r}}|i\rangle\langle i|\boldsymbol{\pi}^- \cdot \hat{\mathbf{r}}|L\rangle$$
$$= \frac{dn(\nu_L)}{d\Omega} \frac{4\pi^2 \alpha}{h} I_i \langle L|\boldsymbol{\pi}^+ \cdot \hat{\mathbf{r}}|i\rangle\langle i|\boldsymbol{\pi}^- \cdot \hat{\mathbf{r}}|L\rangle \tag{12.20}$$

in complete agreement with the result attained in Eq. (12.10). And the result for the differential cross section is the same as given in Eq. (12.11).

At first sight it may seem strange that the same result is achieved using methods based on two quite different assumptions: (i) for excitation of sufficient intensity such that one can neglect spontaneous capture, but for times short compared with the Rabi period; (ii) for weak excitation such that one can neglect stimulated emission. The reason is that each model is effectively a weak excitation one. The condition $\nu_R t \ll 1$ ensures that the probability amplitude of the excited state is always very small. In the first model this comes about because the time is short compared with the Rabi period; in the second because the radiation is weak and therefore ν_R is small. We shall assume that Eq. (12.11) gives the photoelectric differential cross section over a wide range of circumstances.

12.4. THE PHOTOELECTRIC DIFFERENTIAL CROSS SECTION—A MORE GENERAL RESULT

It is possible to obtain a more general expression for the differential cross section by starting with Eq. (12.5) and substituting Eq. (11.30) for the probability of finding the atom in the excited state. In this it is recognized that the "excited state" of the continuum, i.e., the ion produced as the result of photoelectric emission, will decay back to the ground state in a way that is described by a decay constant γ. The probability per unit solid angle that the photoelectron has been produced is

$$\frac{dp(t)}{d\Omega}=\frac{dn(\nu_L)}{d\Omega}\pi^2\nu_R^2$$

$$\times\int_{-\infty}^{+\infty}\left\{\frac{1+e^{-\gamma t}-2\,e^{-\gamma t/2}\cos[(4\pi^2\nu_R^2+4\pi^2\delta^2)^{1/2}t]}{4\pi^2\nu_R^2+\frac{1}{4}\gamma^2+4\pi^2\delta^2}\right\}d\delta \quad (12.21)$$

This may be placed into a more practical form by normalizing according to γ, and writing

$$K=\frac{2\pi\nu_R}{\gamma},\qquad x=\frac{2\pi\delta}{\gamma},\qquad u=\gamma t$$

Then Eq. (12.21) can be written as

$$\frac{dp(t)}{d\Omega}=\frac{dn(\nu_L)}{d\Omega}\frac{\pi^2\nu_R^2}{\gamma}\mathscr{E}_1(K) \quad (12.22)$$

in which the integral has been written as

$$\mathscr{E}_1(K)=\frac{1}{\pi}\int_0^\infty\frac{1+e^{-u}-2\,e^{-u/2}\cos[(K^2+x^2)^{1/2}u]}{K^2+0.25+x^2}dx$$

This integral cannot be expressed in analytical form. Numerical integration for a variety of values of K, which is proportional to the square root of the intensity of the light, have been made. Figure 16a shows the graphs of $\mathscr{E}_1(K)$ as a function of $u=\gamma t$. Several comments need to be made.

When the intensity of the incident light is small, so that $4\pi^2\nu_R^2\ll\gamma^2/4$ and $K\to 0$, the Eq. (12.22) reduces to

$$\frac{dp(t)}{d\Omega}=\frac{dn(\nu_L)}{d\Omega}\frac{\pi^2\nu_R^2}{\gamma}(1-e^{-\gamma t}) \quad (12.23)$$

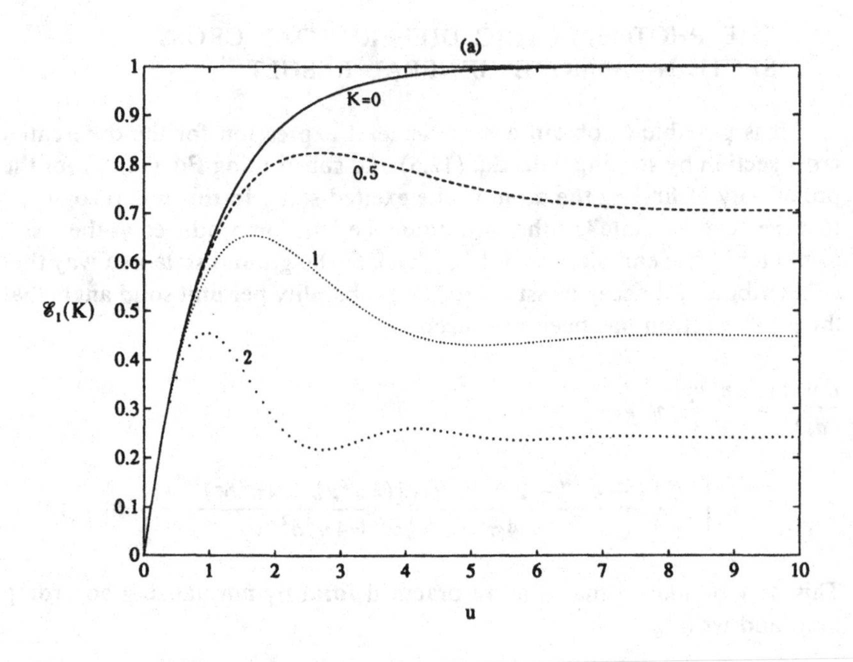

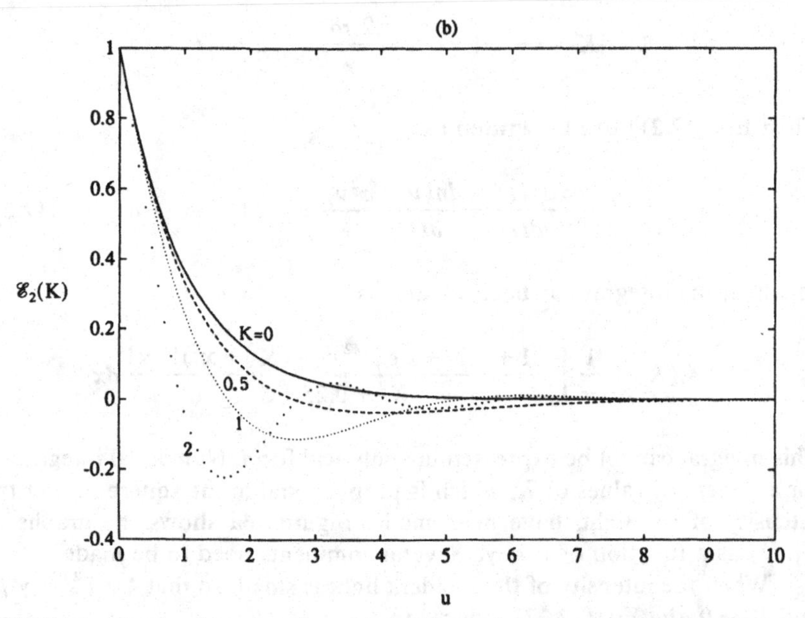

FIGURE 16. Graphs of two functions. (a) $\mathscr{E}_1(K)$ as a function of $u = \gamma t$, proportional to the probability per unit solid angle that photoelectric emission has occurred. (b) $\mathscr{E}_2(K)$ as a function of $u = \gamma t$, proportional to the rate of photoelectric emission. The graphs are drawn for $K = 2\pi\nu_R/\gamma = 0, 0.5, 1, 2$.

spatial and temporal extent; one has constructed, perhaps, a picture of a localized but nonpoint photon. However, on asking the same questions as above, the same sort of difficulties are found. It seems that all models are inadequate, especially when one tries to describe them in classical language.

Let us try to describe what exactly is *observed* in the emission and the absorption processes between two interacting atoms without injecting any model of the interaction process. The experiment to be described is to an extent "gedanken," because I am not sure whether all aspects of the process have been simultaneously observed and measured; but, in principle, one believes that the following could be observed, especially in these days of trapped atoms. The situation is illustrated in Fig. 27. The left of the diagram shows the changes that occur in the source atom S; the right of the diagram shows the changes that occur in the detector atom D; the center of the diagram shows what we presume must be communicated from the source to the detector.

At some instant of time t' the excited source atom makes a transition from its excited state of excitation energy $h\nu_{Sk}$ to a lower or ground level. The atom recoils with momentum $-\mathbf{p}$ and kinetic energy K_S, related by $K_S^2 + 2K_S m_S c^2 = c^2 p^2$. Presumably an amount of energy $h\nu_0 = h\nu_{Sk} - K_S$ with an amount of energy $+\mathbf{p}$ is being transported away by radiation. At some later instant of time t (where $t - t' = R/c$, R being the distance between the source and detector atoms) a detector atom, initially in a ground state, is observed to recoil with momentum $+\mathbf{p}$ and a kinetic K_D, related by $K_D^2 + 2K_D m_D c^2 = c^2 p^2$. The detector atom has made a transition to an excited state of excitation energy $h\nu_{Dk} = h\nu_0 - K_D = h\nu_{Sk} - K_S - K_D$. This process is resonant, and relatively likely to occur, if $h\nu_{Dk}$ is close to one of

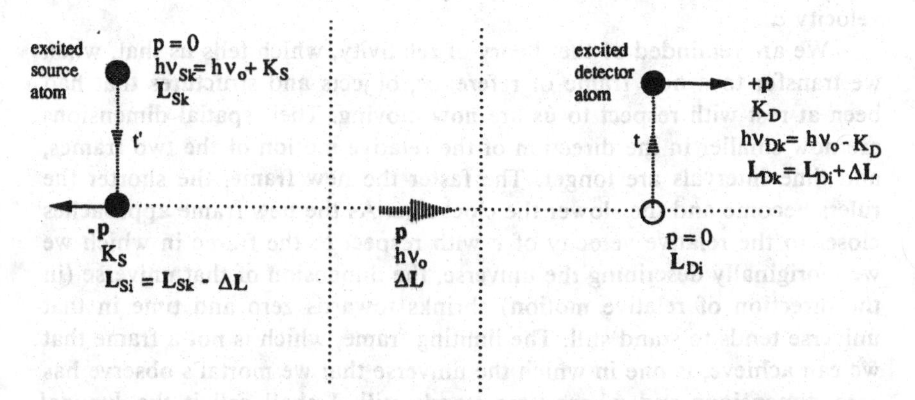

FIGURE 27. Observable changes in the emission-absorption process. The left-hand side of the diagram shows changes in the source atom; the right-hand side shows changes in the detector (absorber) atom.

the allowed excitation states of the detector atom. Of course, the excitation of the detector atom may take it to a state of ionization, in which case **p** is the vector sum of the recoil momentum of the atom and the photoelectron, and K_D is the sum of their kinetic energies. These processes at the detector are also consistent with an amount of energy $h\nu_0$ and a momentum $+\mathbf{p}$ being transported from the source to the detector by radiation.

It is also found that there have been certain changes of angular momenta (in so far as they can be determined under the rules of quantum mechanics) in the source and the detector. Presumably the transport has also involved a transfer of angular momentum given by ΔL in the diagram.

The observations are always in agreement with what is expected under the laws of conservation of energy, linear momentum, and angular momentum. The rules of quantum mechanics correctly describe the changes in the source and detector and the probabilities of the changes occurring. But the question is, how does one describe the interaction process itself? It seems that neither of the above classical-type descriptions, an exchange of a wave or of a particle, will do.

The problem is often referred to as the Einstein–Podolsky–Rosen (EPR) paradox, and the reader is referred to F. Selleri[(1)] for an extensive discussion of the theory and the experimental tests.

14.3. AN ALTERNATIVE DESCRIPTION—THE LUMINAL FRAME

An excited atom decays from an excited state and recoils. Energy and momentum and angular momentum must be transferred to something else in the universe. The transfer is made by a signal that travels at the ultimate velocity c.

We are reminded of the theory of relativity, which tells us that, when we transfer to a new frame of reference, objects and structures that had been at rest with respect to us are now moving. Their spatial dimensions are now smaller in the direction of the relative motion of the two frames, and time intervals are longer. The faster the new frame, the shorter the rulers become and the slower the clocks go. As the new frame approaches closer to the relative velocity of c with respect to the frame in which we were originally describing the universe, the dimension of that universe (in the direction of relative motion) shrinks towards zero and time in that universe tends to stand still. The limiting frame, which is not a frame that we can achieve, is one in which the universe that we mortal's observe has zero dimensions and where time stands still. I shall call it the *luminal frame*—some may prefer to call it *God's frame.* It is the special frame uniquely appropriate to light. Light has this velocity $c = 299\,792\,458 \text{ m s}^{-1}$

with respect to every observer, with respect to every piece of matter, no matter what their velocity, in the universe. But, in the *luminal frame*, light gets everywhere in the universe *instantaneously* (if we can use temporal language); the whole of the universe is in *contact with the source* (if we can use spatial language).

The excited source atom can transfer its energy and angular momentum to another, the detector atom with which it is in contact in the luminal frame, instantaneously. We poor mortals, living in our world, must now wait for a time $t - t' = R/c$ to have the ordained interaction revealed to us. Such a point of view must have religious and theological significance, but that is a subject on which I make no statement.

Of course we must not imagine the whole of the universe in the luminal frame to be without structure—there are directions involved. The source exists in the real world. At the instant it decays we can imagine the immediate start of a spherical electromagnetic wave that is going to get everywhere in the universe instantaneously in the luminal frame. The atom chosen, according to some probability, for the detector is in some direction from the source. Therefore the energy change must take place in a specific direction; there must be equal and opposite momentum changes as well.

Using this picture, we can avoid the usual pitfalls of classical description. Classical description involves the source atom as giving off radiation. The radiation has various descriptions depending on the type of model we prefer. We then try to describe how the detector atom responds to this radiation. The classical description involves the mediation of the field. However, the correct quantum-mechanical description of the process, which gives the correct predictions, describes the source atom and the detector atom as part of a single quantum-mechanical system—they speak directly to one another and not through the intermediary of a light beam but within the rules of electromagnetic interaction. So it is with the description given above using the luminal frame. In this frame the source and detector are in direct or immediate contact; all the decisions about who is to interact with whom, about energy changes, momentum changes, and recoils, about angular momentum changes and polarizations are made instantaneously in that very special frame. We, in our world, must wait for the consequences.

We shall now discuss some experiments using this viewpoint.

14.4. SOURCE–DETECTOR INTERACTION

Let us simplify by assuming that the source atom (S) decays from a $J_k = 1$ state to a $J_i = 0$ ground state. The general expression for the excited state is

$$|S; k\rangle = [a_+ e^{i\alpha_+}|S; 1, +\rangle + a_0|S; 1, 0\rangle + a_- e^{i\alpha_-}|S; 1, -\rangle]\, e^{-i2\pi\nu_k t} \tag{14.2}$$

An unimportant common phase factor has been removed and the remaining amplitudes a_+, a_0, a_- are real. The first label in the kets identifies the atom, the second is the J value, and the third in the M-value +1, 0, −1 in a coordinate system $\mathbf{x}^0$, $\mathbf{y}^0$, $\mathbf{z}^0$ with $\mathbf{z}^0$ being the axis of quantization. The ground state of the atom is $|S; i\rangle = |S; 0, 0\rangle$.

The dipole moment matrix element describing the transition is

$$\langle S; i|\hat{\mathbf{D}}|S; k\rangle$$

$$= \left(\frac{\mathbf{x}^0 + i\mathbf{y}^0}{\sqrt{2}}\, a_+\, e^{i\alpha_+} + \mathbf{z}^0 a_0 + \frac{\mathbf{x}^0 - i\mathbf{y}^0}{\sqrt{2}}\, a_-\, e^{i\alpha_-}\right) \frac{D_S}{\sqrt{3}}\, e^{-i2\pi\nu_0' t'} \tag{14.3}$$

D_S is the reduced matrix element of the dipole transition and the factor $1/\sqrt{3}$ and the unit (complex) vectors come from the analysis of Eqs. (7.23)–(7.26)—for the $J = 1$ to $J = 0$ transition all components have the same strength. The recoil of the ground state atom has been neglected but some recognition is given in the replacement of ν_k by ν_0'. The prime on t' indicates that this is the time at which the transition in the source atom takes place.

In the laboratory this transition generates an electromagnetic wave. At a field position $\mathbf{r}$ relative to the source this creates an electric vector derived from

$$\mathbf{E}^+(\mathbf{r}, t) = \frac{E_0}{\sqrt{2}}\left(\frac{\mathbf{x}^0 + i\mathbf{y}^0}{\sqrt{2}}\, a_+\, e^{i\alpha_+} + \mathbf{z}^0 a_0 + \frac{\mathbf{x}^0 - i\mathbf{y}^0}{\sqrt{2}}\, a_-\, e^{i\alpha_-}\right) \cdot$$

$$(\boldsymbol{\theta}^0\boldsymbol{\theta}^0 + \boldsymbol{\phi}^0\boldsymbol{\phi}^0)\, e^{-i2\pi\nu_0' t} \tag{14.4}$$

which involves the projection of the vector character of the dipole onto the $\boldsymbol{\theta}^0\boldsymbol{\phi}^0$ plane transverse to $\mathbf{r}^0$. The rms amplitude E_0 is related to the reduced matrix element by the factors given in Eq. (3.34) and the correspondence principle, Eq. (7.43). Using the projections

$$\begin{aligned} \mathbf{x}^0 \cdot \boldsymbol{\theta}^0 &= \cos\theta \cos\phi, & \mathbf{x}^0 \cdot \boldsymbol{\phi}^0 &= -\sin\phi \\ \mathbf{y}^0 \cdot \boldsymbol{\theta}^0 &= \cos\theta \sin\phi, & \mathbf{y}^0 \cdot \boldsymbol{\phi}^0 &= \cos\phi \\ \mathbf{z}^0 \cdot \boldsymbol{\theta}^0 &= -\sin\theta, & \mathbf{z}^0 \cdot \boldsymbol{\phi}^0 &= 0 \end{aligned} \tag{14.5}$$

where θ and ϕ are the polar and azimuthal angles of $\mathbf{r}$ in $\mathbf{x}^0$, $\mathbf{y}^0$, $\mathbf{z}^0$, we obtain

$$\mathbf{E}^+(\mathbf{r}, t) = \frac{E_0}{\sqrt{2}}\left(\frac{\boldsymbol{\theta}^0 \cos\theta + i\boldsymbol{\phi}^0}{\sqrt{2}}\, e^{i\phi}\, e^{i\alpha_+} a_+ + \boldsymbol{\theta}^0 \sin\theta a_0\right.$$

$$\left. + \frac{\boldsymbol{\theta}^0 \cos\theta - i\boldsymbol{\phi}^0}{\sqrt{2}}\, e^{-i\phi}\, e^{i\alpha_-} a_-\right) e^{-i2\pi\nu_0 t} \tag{14.6}$$

It can immediately be seen that part of the information about the original excitation has been lost; the electric field has been projected onto a plane whereas the dipole moment was in three dimensions. If we allow the electric field of Eq. (14.6) to interact with the detector atom (D) we shall obtain an incomplete description of its excited state; some component of angular momentum has been lost from the system. If we take the case of a detector atom with the same structure as the source (resonance absorption) and allow *direct interaction*, as we have deemed possible using the luminal frame, the excited state will be described by

$$|D; k\rangle = (a_+ e^{i\alpha_+}|D; 1, +\rangle + a_0|D; 1, 0\rangle + a_- e^{i\alpha_-}|D; 1, -\rangle)\, e^{-i2\pi\nu_0 t} \qquad (14.7)$$

with identical amplitudes and phases to those in the excited state of the source atom. To allow for the effects of momentum conservation the excitation frequency has been written as $\nu_0 = \nu_k -$ (the recoil kinetic energies). Of course, if the radiation from the source atom passes through a polarizer before interacting with the detector, that polarizer forms part of the detector and some of the information is lost to the detector atom (although, intrinsically, the missing information is in the reaction of the polarizer and the complete detector system, polarizer plus detector atom, contains the full information).

This case, the decay and recoil of a source atom and the excitation of a detector atom, will not be discussed further. But the consequences of the viewpoint developed will be used in the following discussion of the experiments to study the physics of cascade transitions.

14.5. CASCADE OPTICAL TRANSITIONS

Consider an atom that decays from an excited $J_k = 0$ state, through an intermediate state with $J_j = 1$, to a final (ground) state with $J_i = 0$. The first transition will be labeled A with a frequency ν_A, and the second will be labeled B with a frequency ν_B. The structure of the atomic system is illustrated in Fig. 28a; the energies of the three levels of the intermediate state are taken to be degenerate. In a typical experiment to study the polarization correlations between these cascade transitions two detectors are placed on either side of the source (Fig. 28b). (See, for example, the experiments of A. Aspect *et al.*[2] and the extensive discussions given in Ref. 1. A very entertaining and readable account of the experiment and its problems of interpretation is given by N. D. Mermin.[3]) The detector D_A

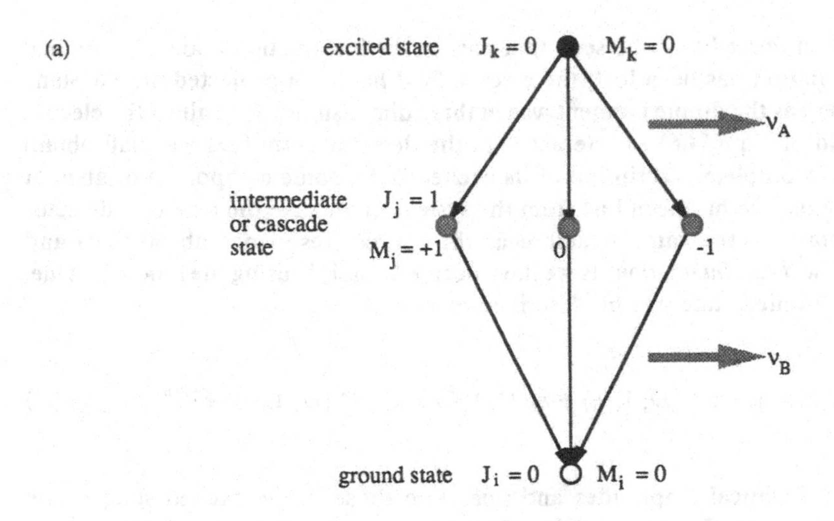

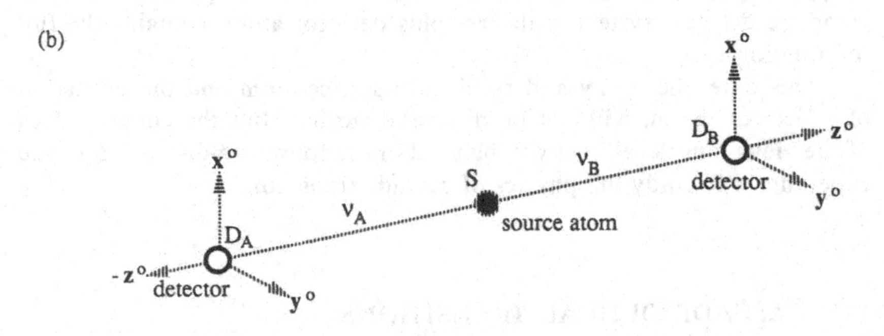

FIGURE 28. (a) The structure of the three-level atom in cascade transitions. (b) An experimental arrangement to detect the cascade transitions A and B.

in the $-\mathbf{z}^0$ direction is tuned to respond only to ν_A; the detector D_B in the $+\mathbf{z}^0$ direction is tuned to respond only to ν_B. Using this z axis as the axis of quantization for the description of the atom, the three states are

$$|k\rangle = |k; 0, 0\rangle\, e^{-i2\pi(\nu_A+\nu_B)t}$$

$$|j\rangle = (a_+\, e^{i\alpha_+}|j; 1, +\rangle + a_0|j; 1, 0\rangle + a_-\, e^{i\alpha_-}|j; 1, -\rangle)\, e^{-i2\pi\nu_B t} \tag{14.8}$$

$$|i\rangle = |i; 0, 0\rangle$$

Since the detection is along the $\pm z$ axis the $M = 0$ state of the intermediate level does not contribute, so we can redefine the states, using now the state $|j\rangle$ as the zero of energy:

$$|k\rangle = |k; 0, 0\rangle\, e^{-i2\pi\nu_A t}$$

$$|j\rangle = a_+\, e^{-i\delta}|j; 1, +\rangle + a_-\, e^{+i\delta}|j; 1, -\rangle \tag{14.9}$$

$$|i\rangle = |i; 0, 0\rangle\, e^{+i2\pi\nu_B t}$$

where we have written $\alpha_\pm = \alpha \mp \delta$ and have suppressed the arbitrary phase α.

The A transition at time t_A is described by the matrix element of the dipole moment:

$$\langle j|\hat{\mathbf{D}}|k\rangle = \left(\frac{\mathbf{x}^0 - i\mathbf{y}^0}{\sqrt{2}}\, a_+\, e^{i\delta} + \frac{\mathbf{x}^0 + i\mathbf{y}^0}{\sqrt{2}}\, a_-\, e^{-i\delta}\right)\frac{D_A}{\sqrt{3}}\, e^{-i2\pi\nu_A t_A} \tag{14.10}$$

The B transition at time t_B is described by the matrix element of the dipole moment:

$$\langle i|\hat{\mathbf{D}}|j\rangle = \left(\frac{\mathbf{x}^0 + i\mathbf{y}^0}{\sqrt{2}}\, a_+\, e^{-i\delta} + \frac{\mathbf{x}^0 - i\mathbf{y}^0}{\sqrt{2}}\, a_-\, e^{i\delta}\right)\frac{D_B}{\sqrt{3}}\, e^{i2\pi\nu_B t_B} \tag{14.11}$$

The polarization character of the radiations depend on the cascade path and therefore on the values of a_+, a_-, and δ. For example, if the cascade travels through the $M = +1$ state of $|j\rangle$, $a_+ = 1$, $a_- = 0$. The A dipole is circularly polarized as shown in Fig. 29a, rotating at frequency ν_A from $\mathbf{x}^0$ to $-\mathbf{y}^0$; the radiation towards the detector D_A is traveling along $-\mathbf{z}^0$ and has right-hand helicity. The B dipole is circularly polarized in the opposite rotation sense, but, since the radiation to be detected is traveling in the opposite direction, $+\mathbf{z}^0$, it also has right-hand helicity.

We shall, however, discuss the case where linear polarizers are to be used in front of the detectors. Linearly polarized dipoles are produced when $a_+ = a_- = 1/\sqrt{2}$ (neglecting the path through $M = 0$ because it cannot be detected with the chosen orientations and the polarizations used). In this case we have

$$|j\rangle = \frac{1}{\sqrt{2}}\left(e^{-i\delta}|j; 1, +\rangle + e^{i\delta}|j; 1, -\rangle\right) \tag{14.12}$$

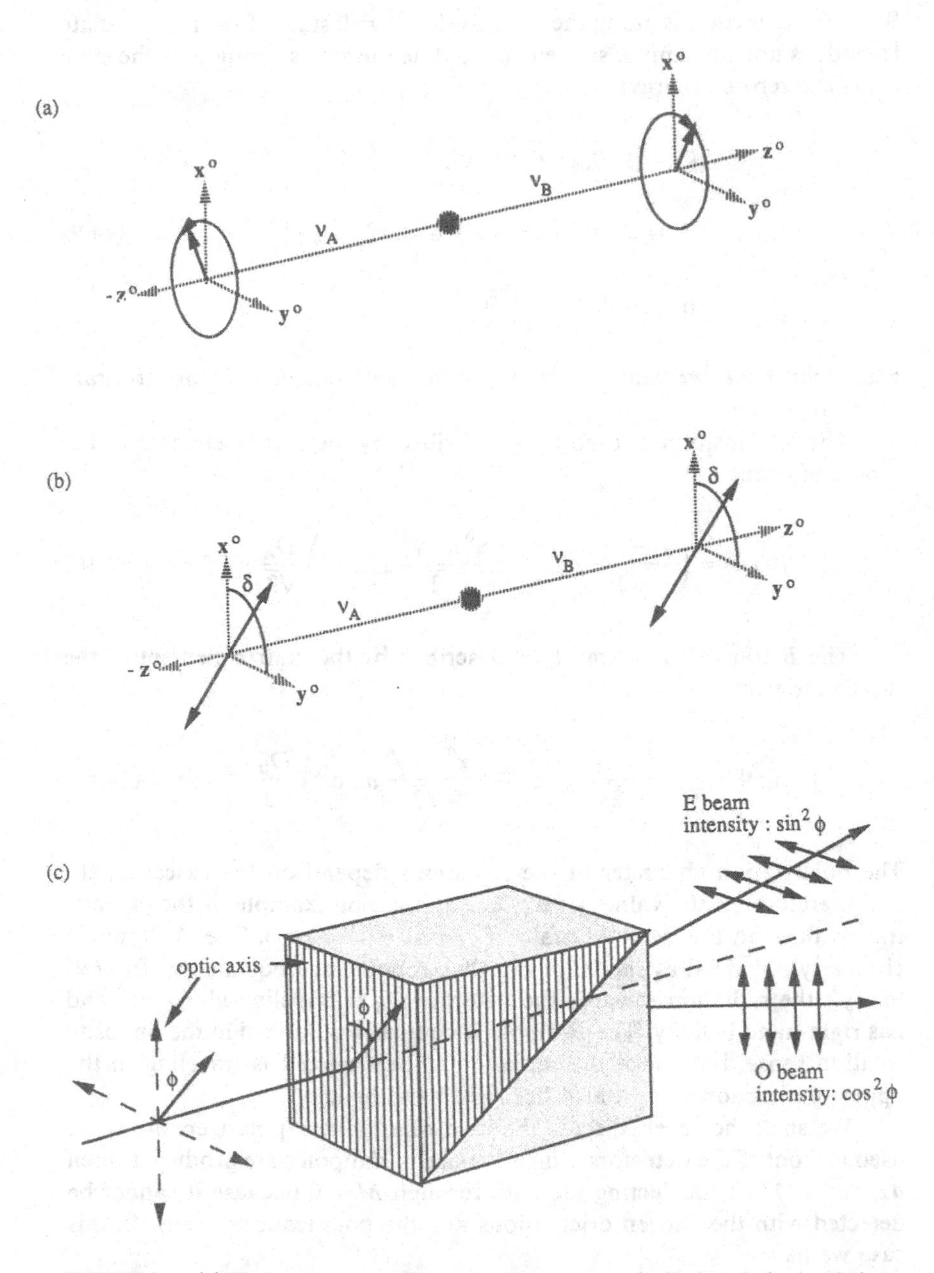

FIGURE 29. (a) The polarization relationships between two circularly polarized emissions in a cascade. (b) The polarization relationships between two linearly polarized emissions in a cascade. (c) A Wollaston prism type analyzer for linearly polarized radiations.

and the dipole matrix elements of Eqs. (14.10) and (14.11) become

$$\langle j|\hat{\mathbf{D}}|k\rangle = [\mathbf{x}^0 \cos\delta + \mathbf{y}^0 \sin\delta](2/3)^{1/2} D_A\, e^{-i2\pi\nu_A t_A}$$
$$\langle i|\hat{\mathbf{D}}|j\rangle = [\mathbf{x}^0 \cos\delta + \mathbf{y}^0 \sin\delta](2/3)^{1/2} D_B\, e^{i2\pi\nu_B t_B} \qquad (14.13)$$

It is seen that, for this geometry, the vector character of each dipole and the polarization of each beam are the same, linear at an angle δ to the x axis (Fig. 29b). In front of each detector a polarization analyzing prism is placed. This is provided, for example, by a Wollaston prism illustrated in Fig. 29c. The incident light is split into two beams—the ordinary beam (O) being that part of the incident light polarized parallel to the optic axis of the first part of the prism, and the extraordinary beam (E) being that part of the incident light polarized parallel to the optic axis of the second part of the prism. Thus, if the first optic axis is at an angle ϕ to the polarization direction of the light, the probability of light entering the O beam is $\cos^2\phi$ and the probability of it entering the E beam is $\sin^2\phi$. The experiment is performed at sufficiently low excitation rates that a single cascade is observed, ν_A at D_A followed by ν_B at D_B. The delay between the two counts is determined by the mean life of the intermediate level, but the experiment is performed so that only those counts at A that are in coincidence with a count at B are recorded. The experiment is performed by arranging that for the chosen angle for $\phi_A = 0°, 120°, 240°$, the orientation of the optic axis of D_A, the optic axis of D_B is placed at $\phi_B = \phi_A, \phi_A + 120°, \phi_A + 240°$. The experiment is repeated a large number of times with a random but equal selection of ϕ_A and ϕ_B, and for each it is recorded whether the count is received in the O channel or the E channel. Thus, for each record there are four possible outcomes, *OO, OE, EO, EE,* being the channels in which a count is recorded for the A beam and the B beam respectively.

Let D_A be set at $\phi_A = 0°$, ($\mathbf{x}^0$ direction). Suppose a count is received in the O channel of D_A. Using the argument above regarding the direct interaction between the source atom and the detector, we can immediately say that the source must be left in the state with $\delta = 0°$. From Eq. (14.12) the atom *must* have been left in the state

$$|j\rangle = \frac{1}{\sqrt{2}}(|j; 1, 0\rangle + |j; 1, -\rangle)$$

This state subsequently decays to ground with a matrix element

$$\langle i|\hat{\mathbf{D}}|j\rangle = \mathbf{x}^0(\tfrac{2}{3})^{1/2} D_B\, e^{i2\pi\nu_B t_B}$$

If the detector D_B is oriented at $\phi_B = 0°$, i.e., along $\mathbf{x}^0$, the count *must* be received into the O channel.

On the other hand, if the first count had been received in the E channel of D_A, then the source atom must be left in the state with $\delta = 90°$. The intermediate state must be

$$|j\rangle = \frac{-i}{\sqrt{2}}(|j; 1, +\rangle - |j; 1, -\rangle)$$

This state is polarized in the y^0 direction and the subsequent ν_B radiation *must* go into the E channel of D_B. Thus, if the orientations of D_A and D_B are the same, the only possible outcomes are OO or EE.

Suppose, however, that, with $\phi_A = 0°$, we orient D_B at $\phi_B = 120°$. The source has been left in the state with $\delta = 0°$. The probability of the count being recorded in channel O of D_B is $\cos^2 120° = 1/4$ and the probability of it being recorded in channel E is $\sin^2 120° = 3/4$; the same is the case for $\phi_B = 240°$. We can draw up a table of probabilities for the nine different choices for ϕ_A and ϕ_B, and for the four different outcomes (Table I). The table is normalized to a total of 72 trials. The overall probability of any particular result OO, OE, EO, EE is the same—18 out of 72. The significant feature of this probability table is that, if the two detectors have the same orientation, the only outcomes possible are OO or EE; i.e., *it would appear that the second detector knows how the first had been oriented.* The resolution of this so-called paradox does not need superluminal velocities to communicate between the detectors nor the concept of collapsing wave packets. It is only necessary to realize that, because of the concept of *luminal frames*, the outcome of any sequence of events as we observe them is preordained.

TABLE I
Aspect Experiments: Relative Probabilities of the Four Different Outcomes (OO, OE, EO, EE) for Nine Different Settings of the Detector Orientations

Settings		Outcomes			
ϕ_A	ϕ_B	OO	OE	EO	EE
0°	0°	4	0	0	4
0°	120°	1	3	3	1
0°	240°	1	3	3	1
120°	0°	1	3	3	1
120°	120°	4	0	0	4
120°	240°	1	3	3	1
240°	0°	1	3	3	1
240°	120°	1	3	3	1
240°	240°	4	0	0	4

This picture gives a rationale for the quantum-mechanical concept of "the whole system."

Before leaving the discussion of this important experiment, it is worthwhile looking at the situation when circular polarizers are used in the detection process. Circular polarization would be analyzed by placing a quarter-wave plate in the beam. Figure 30a illustrates how right-hand and left-hand helicity light are turned to linearly polarized of orthogonal directions. A quarter-wave plate transforms light by the action of the operator $(\mathbf{ii} + i\mathbf{jj})\cdot$ where $\mathbf{i}$ is the unit vector in the direction of the fast axis and $\mathbf{j}$ is the unit vector in the direction of the slow axis.

A circularly polarized dipole described by the polarization vector $(\mathbf{x}^0 + i\mathbf{y}^0)/\sqrt{2}$ produces radiation in the $\mathbf{r}^0$ (i.e., θ, ϕ direction) with polarization character

$$(\boldsymbol{\theta}^0\boldsymbol{\theta}^0 + \boldsymbol{\phi}^0\boldsymbol{\phi}^0)\cdot\frac{\mathbf{x}^0 + i\mathbf{y}^0}{\sqrt{2}} = \frac{\boldsymbol{\theta}^0\cos\theta + i\boldsymbol{\phi}^0}{\sqrt{2}}\,e^{i\phi}$$

$$= \left(\frac{\cos\theta + 1}{2}\,\frac{\boldsymbol{\theta}^0 + i\boldsymbol{\phi}^0}{\sqrt{2}} + \frac{\cos\theta - 1}{2}\,\frac{\boldsymbol{\theta}^0 - i\boldsymbol{\phi}^0}{\sqrt{2}}\right)e^{i\phi}$$

The elliptically polarized light can be regarded as a superposition of two circularly polarized beams. This elliptically polarized light becomes transformed to a superposition of othogonally linearly polarized beams on passage through a quarter-wave plate:

$$(\mathbf{ii} + i\mathbf{jj})\cdot\frac{\boldsymbol{\theta}^0\cos\theta + i\boldsymbol{\phi}^0}{\sqrt{2}}\,e^{i\phi}$$

$$= \frac{\cos\theta + 1}{2}\,\frac{\mathbf{i} - \mathbf{j}}{\sqrt{2}}\,e^{i\alpha}\,e^{i\phi} + \frac{\cos\theta - 1}{2}\,\frac{\mathbf{i} + \mathbf{j}}{\sqrt{2}}\,e^{-i\alpha}\,e^{i\phi}$$

where α is the angle between the $\boldsymbol{\theta}^0$ direction used to describe the circularly polarized light and the $\mathbf{i}$ fast axis of the quarter-wave plate. The light has thereby been resolved into two orthogonal channels, one of relative intensity $a^2 = [(\cos\theta + 1)/2]^2$ for the right-hand circular component and the other of relative intensity $b^2 = [(\cos\theta - 1)/2]^2$ for the left-hand circular component.

The equivalent experiment to that described above but using circularly polarized light would be performed by placing the three D_B detectors around a circle at $\theta = 0°$, 120°, 240°, and $\phi = 0°$ as shown in Fig. 30b. Thus the

(a)

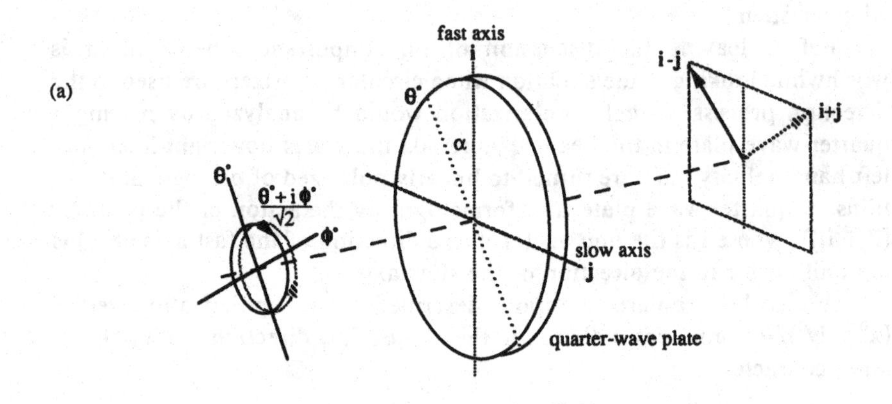

(b)

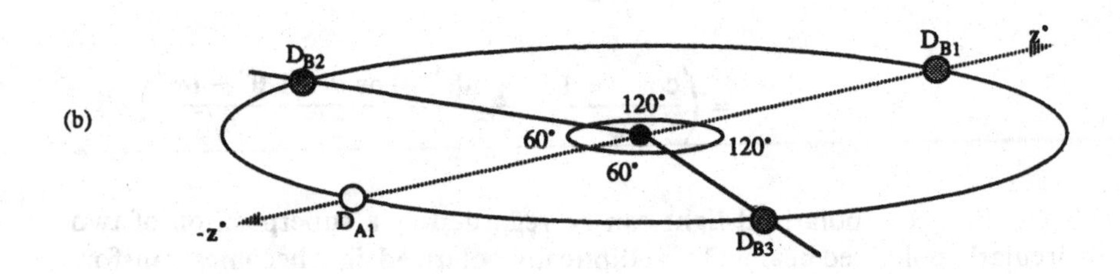

(c)

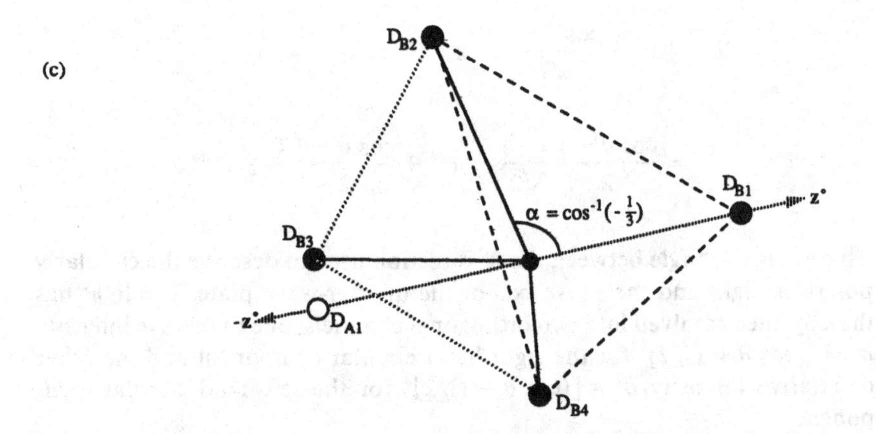

FIGURE 30. (a) The action of a quarter-wave plate analyzer. (b) The Aspect experiment using circularly polarized light. (c) The experiment in three dimensions.

TABLE II

Settings: $\theta_A = 0°$		Outcomes			
$\cos\theta_B$	ϕ_B	*OO*	*OE*	*EO*	*EE*
1	0°	9	0	0	9
$-\frac{1}{3}$	0°	1	4	4	1
$-\frac{1}{3}$	120°	1	4	4	1
$-\frac{1}{3}$	240°	1	4	4	1

intensities of the O and E components, proportional to a^2 and b^2, will be

$$\text{for } \theta = 0°: \quad a^2 = 1 \text{ and } b^2 = 0$$

$$\text{for } \theta = 120°, 240°: \quad a^2 = 1/4 \text{ and } b^2 = 3/4$$

exactly the same probabilities as for Table I above.

This two-dimensional array could be extended to three dimensions as shown in Fig. 30c. For each position of D_A, there are four positions for the detectors D_B, at the corners of a tetrahedron. For this geometry, the position 1 has $\theta = 0°$, while the positions 2, 3, 4 have $\theta_B = \cos^{-1}(-1/3)$, $\phi_B = 0°$, 120°, 240°. The first quarter of the probability table would now read (normalized to a total of 264 counts over the four different positions for D_A) as shown in Table II.

Thus, by using the quantum mechanical predictions of the probabilities of certain outcomes, which can be supported by using the model of the luminal frame, the observed results are obtained. This contrasts with what would have been incorrectly predicted if one used a model of polarized beams of light or polarized photons carrying the information out from the source to the detectors.

14.6. THE SAME EXPERIMENT WITH PARTICLES

It is believed that the same result, as expressed in the probability table above, would be achieved with particles instead of light beams. This can be done, for example, by creating two spin-$\frac{1}{2}$ particles in the triplet state (i.e., with parallel spins) flying apart in opposite directions towards the two detectors, such as are produced in the production of a positive and negative electron pair created from a photon of energy greater than $2m_ec^2$ in the field of a nucleus. Each detector now consists of a Stern–Gerlach magnet which separates the two opposite states of electron spin into two channels for detection (Fig. 31). Each magnet, with its counting equipment forming D_A and D_B, can be oriented at the azimuthal angle of 0°, 120°, 240°. The

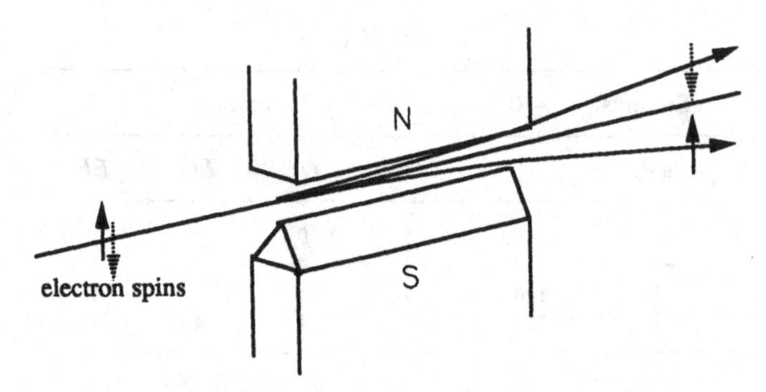

FIGURE 31. A Stern–Gerlach magnet for separating electrons with different spin components. The magnetic moment of the electron is directed opposite to its spin. If the magnetic moment is directed along the field direction it will be forced towards the region of stronger field.

quantum mechanical description of the process will give the same results as above. It may be felt that, since the two electrons travel, of necessity, at less than the speed of light, the arguments used about the luminal frame are no longer appropriate. However, consider the following.

Right from the moment of its production a charged particle, or indeed a neutral particle with magnetic moment, must be in communication with the rest of the universe. It is the state of charge of the universe and the currents created by the motion of its charged particles that establish the electromagnetic field, static or dynamic, in which that particle must immediately move. Now, as pointed out in Section 14.1, a field is only *our way* of modeling the interaction between a source and a test charge. It is an open philosophical question whether the field actually exists if there is no test charge wanting to know what is interaction with the rest of the universe must be. Taking this point of view, a charge, at its moment of creation, establishes contact, by electromagnetic interaction in the luminal frame, with every other source in the universe and, thereby, finds out how it is to move and what atomic system it is "eventually" (i.e., in the frame of us mortals) to interact with and with what result. This is, presumably, connected with the fact that the static field vectors at a field point P due to a moving charge are determined by the *presumed position* of the source charge *at the time* t, not by the retarded position at the retarded time t', as mentioned in Chapter 2.

14.7. THE ANSWER

"Light,"—in which term we include the whole of the electromagnetic spectrum of radiation—is the word we use to describe the interaction between the motion of charges in the sources and those in the detectors.

The atoms and charged particles in the sun, and other sources, interact with the atoms, etc. in the retina of our eyes. We say "light has traveled from the sun to our eyes"; and indeed we can measure the delay time for that travel and establish the speed of that light. However, the nature of the process, just as for the other type of electromagnetic interaction by static fields, is itself unknown. To ask for its nature is not really a valid question; the nature of light is the sum of its properties, and is not specified by any model we may wish to erect for it. Nonetheless, it is in our nature to want to picture the process, to want to have a language with which we can describe what is observed, without having to resort only to mathematical formalism. Even the mathematician and theoretical physicist need a language. The language of waves of electric and magnetic fields, quantities already defined for static interactions, enables us to describe successfully how the waves are propagated through space. When the quantum rules for interaction are added we get a satisfactory explanation of what happens in emission, absorption, excitation, scattering processes. And finally, if we use the concept of the luminal frame we can avoid the "spooky" aspects of action at a distance.

In the luminal frame there is no need for words like "action at a distance," "collapse of wave function," "superluminal velocities," "advanced potentials," etc.—the language invented to explain observations that appear strange. The concept of advanced potentials as used in absorber theory (see Ref. 4) is, in fact, the equivalent of the luminal frame description, but expressed in the words and frame of the laboratory observer. In the luminal frame everything in the system is in immediate contact.

To bathe objects in light is to merge them with the infinite

—Leonardo da Vinci[5]

REFERENCES

1. *Quantum Mechanics versus Local Realism*, F. SELLERI (Ed.) (Plenum Press, New York, 1988).
2. A. ASPECT, P. GRANGIER, J. DALIBARD, AND G. ROGER, see the review "Experimental Investigation of the Einstein-Podolsky-Rosen Question and Bell's Inequality," by A. J. Duncan and H. Kleinpoppen in Ref. 1.
3. N. D. MERMIN, "Is the Moon there when Nobody Looks? Reality and the Quantum Theory," *Physics Today* **38**, 38–47 (April 1985).
4. D. T. PEGG, "Absorber Theory in Quantum Optics," *Phys. Scr.* **T12**, 14–18 (1986).
5. L. BORTOLON (translated by C. J. Richards), *The Life and Times of Leonardo* (Paul Hamlyn, London, 1968), p. 21.

APPENDIX 1

Time Averaging

In the treatment of problems involving a polychromatic radiation field one frequently meets with a real physical quantity that is the square of a physical quantity, itself a function of time:

$$X(t) = CF(t)F(t) \tag{A1.1}$$

Examples occur in Eq. (3.21), the radiant power from an oscillator, and in Eq. (5.2) for the energy density of a field. In practice it is not $X(t)$ that is measured, but some time average over an interval of time T. This is written as $\langle X(t)\rangle_T$ when the interval T is finite, in which case some residual time fluctuation of lower frequencies, $f < 1/T$, remain; or it is written as $\langle X(t)\rangle$ is T is infinitely long so that a true stready-state has been obtained; or indeed simply as X if it is intrinsically independent of time (such as, for example, in the energy density of a circularly polarized monochromatic wave).

In general, $F(t)$ is expanded into a frequency spectrum by Fourier transformation:

$$F(t) = \int_{-\infty}^{+\infty} d\nu\, \tilde{F}(\nu)\, e^{-i2\pi\nu t} \tag{A1.2}$$

$\tilde{F}(\nu)$ is the Fourier transform of $F(t)$. Equation (A1.1) becomes

$$X(t) = C\int_{-\infty}^{+\infty} d\nu\, \tilde{F}(\nu)\, e^{-i2\pi\nu t} \int_{-\infty}^{+\infty} d\nu''\, \tilde{F}(\nu'')\, e^{-i2\pi\nu'' t}$$

Place $\nu'' = -\nu'$ and use Eq. (3.18), $\tilde{F}(-\nu') = \tilde{F}^*(\nu')$:

$$X(t) = C\int_{-\infty}^{+\infty} d\nu\, \tilde{F}(\nu) \int_{-\infty}^{+\infty} d\nu'\, \tilde{F}^*(\nu')\, e^{i2\pi(\nu'-\nu)t} \tag{A1.3}$$

We are now ready to perform time averaging. We consider two forms of time averaging:

A1.1. TRUNCATED TIME AVERAGING

$X(t)$ is averaged over all times in the interval $t - T < t' < t$, with equal weighting throughout the interval T:

$$\langle X(t)\rangle_T = \frac{1}{T}\int_{t-T}^{t} dt'\, C \int_{-\infty}^{+\infty} d\nu\, \tilde{F}(\nu) \int_{-\infty}^{+\infty} d\nu'\, \tilde{F}^*(\nu')\, e^{i2\pi(\nu'-\nu)t'}$$

$$= \int_{-\infty}^{+\infty} d\nu \int_{-\infty}^{+\infty} d\nu'\, C\tilde{F}(\nu)\tilde{F}^*(\nu') \frac{1}{T}\int_{t-T}^{t} dt'\, e^{i2\pi(\nu'-\nu)t'} \tag{A1.4}$$

The integration will normally be carried out over a time interval T which is very much greater than the typical signal period $1/\nu$. In this case the final integral over time in Eq. (A1.4) is only significant for frequencies $\nu' \approx \nu$. Therefore, we place $\nu' = \nu + f$, where f is a small frequency. Then, after carrying out the time integration, we obtain

$$\langle X(t)\rangle_T = \int_{-\infty}^{+\infty} d\nu \int_{-\infty}^{+\infty} df\, C\tilde{F}(\nu)\tilde{F}^*(\nu + f)\, e^{i2\pi ft}\, \mathscr{F}(f, T)\, e^{-i\pi fT} \tag{A1.5}$$

The factor $\mathscr{F}(f, T)$ is the Fraunhofer factor defined in Appendix 5.3 [Eq. (A5.23)]:

$$\mathscr{F}(f, T) = \frac{\sin(\pi fT)}{\pi fT} \tag{A1.6}$$

Graphs of $\mathscr{F}(f, T)$ and the phase πfT are shown in Fig. 33 (Appendix 5). Note that, as demonstrated in Eqs. (A5.24) and (A5.26), as $T \to \infty$, both $\mathscr{F}(2f, T)$ and $\mathscr{F}(f, T)\, e^{-i\pi fT}$ behave as $(1/2T)\delta(f)$.

At this point we shall consider the case when $T \to \infty$, leaving until later consideration of aspects of the situation where T remains finite. Using the δ function, Eq. (A1.5) becomes

$$\langle X\rangle = \frac{1}{2}\int_{-\infty}^{+\infty} d\nu \lim_{T\to\infty} \frac{C\tilde{F}(\nu)\tilde{F}^*(\nu)}{T} \tag{A1.7}$$

Because the integrand is real, and an even function of ν, this can be written as an integral over positive frequencies only:

$$\langle X\rangle = \int_{0}^{\infty} d\nu \lim_{T\to\infty} \frac{C\tilde{F}(\nu)\tilde{F}^*(\nu)}{T} \tag{A1.8}$$

The spectral distribution of the field X is therefore given by

$$\rho(\nu) = \lim_{T\to\infty} \frac{C\tilde{F}(\nu)\tilde{F}^*(\nu)}{T} \tag{A1.9}$$

A1.2. EXPONENTIAL AVERAGING

It is difficult to imagine a physical mechanism for accomplishing the truncated averaging discussed in the last section. A more physically realizable process would use exponential averaging, as occurs, for example, in a photodetector whose output current is measured in a circuit with an integrating time constant $T = RC$; the signal at time t has memory about information at all earlier times $t' < t$ with a weighting of $(1/T)\, e^{-(t-t')}$. The expression for the average is now

$$\langle X\rangle_T = \int_{-\infty}^{+\infty} d\nu \int_{-\infty}^{+\infty} d\nu'\, C\tilde{F}(\nu)\tilde{F}^*(\nu')$$

$$\times \frac{1}{T}\int_{-\infty}^{t} dt'\, e^{i2\pi(\nu'-\nu)t'}\, e^{+(t'-t)/T} \tag{A1.10}$$

After placing $\nu' = \nu + f$ and carrying out the time integration, we have

$$\langle X\rangle_T = \int_{-\infty}^{+\infty} d\nu \int_{-\infty}^{+\infty} df\, C\tilde{F}(\nu)\tilde{F}^*(\nu+f)\, e^{i2\pi ft}\, \mathscr{L}(f, T)\, e^{-i\lambda(f)} \tag{A1.11}$$

The Lorentz factor and its associated phase are given by (see Appendix 5, Section A5.2):

$$\mathscr{L}(f, T) = \frac{1}{(1+4\pi^2 f^2 T^2)^{1/2}}$$

$$\lambda(f) = \arctan(2\pi f T) \tag{A1.12}$$

Graphs of the modulus and phase are shown in Fig. 32 (Appendix 5). Note that, as shown in Eqs. (A5.15) and (A5.19), as $T \to \infty$, both $\mathscr{L}^2(f, T)$ and $\mathscr{L}(f, T)\, e^{-i\lambda(f)}$ behave as $(1/2T)\delta(f)$. Therefore, in the case of $T \to \infty$, the values of $\langle X\rangle$ and of its spectral distribution $\rho(\nu)$ are the same as in Eqs. (A1.8) and (A1.9). In other circumstances, when T is finite, the details will be slightly different. When dealing with specific cases from here on, exponential averaging will be used.

A1.3. AVERAGING OVER A FINITE TIME INTERVAL

When the averaging period T is finite, we must study further the expressions for $\langle X(t)\rangle_T$ given by Eq. (A1.11) and, in particular, the effect of the averaging factor $\mathcal{L}(f, T)\, e^{-i\lambda(f)}$. In order to express Eq. (A1.11) in terms of positive signal frequencies, it can be written, using Eq. (3.18), as

$$\langle X\rangle_T = \int_0^{+\infty} d\nu \int_{-\infty}^{+\infty} df\, C[\tilde{F}(\nu)\tilde{F}^*(\nu+f) + \tilde{F}^*(\nu)\tilde{F}(\nu-f)]$$
$$\times\, e^{i2\pi f t}\mathcal{L}(f, T)\, e^{-i\lambda(f)} \qquad \text{(A1.13)}$$

The spectral distribution is therefore

$$\langle \rho(\nu)\rangle_T = \int_{-\infty}^{+\infty} df\, C[\tilde{F}(\nu)\tilde{F}^*(\nu+f) + \tilde{F}^*(\nu)\tilde{F}(\nu-f)]$$
$$\times\, e^{i2\pi f t}\mathcal{L}(f, T)\, e^{-i\lambda(f)} \qquad \text{(A1.14)}$$

If T is not infinite, the finite spectral width of the averaging function $\mathcal{L}(f, T)\, e^{-i\lambda(f)}$ means that some low frequencies are retained in the signal, those with $f < 1/T$ being the most significant. These frequencies are caused by the beating together of signal frequencies ν and $\nu' = \nu \pm f$ in the field that contributes to the product $X(t)$ in Eq. (A1.1).

We now apply these results to several important cases.

A1.4. THE ENERGY DENSITY AND SPECTRAL DISTRIBUTION OF "BROAD-BAND" OR "WHITE" LIGHT

With $C = \varepsilon_0 = 1/\mu_0 c^2$ and $F(t) = \mathbf{E}(t)$, Eq. (A1.1) becomes the expression for the energy density of the electromagnetic field—cf. Eq. (5.2):

$$u(t) = \varepsilon_0 \mathbf{E}(t)\cdot\mathbf{E}(t)$$

Equation (A1.14) then becomes

$$\langle \rho(\nu)\rangle_T = \int_{-\infty}^{+\infty} df\, \varepsilon_0[\tilde{\mathbf{E}}(\nu)\cdot\tilde{\mathbf{E}}^*(\nu+f) + \tilde{\mathbf{E}}^*(\nu)\cdot\tilde{\mathbf{E}}(\nu-f)]$$
$$\times\, e^{i2\pi f t}\mathcal{L}(f, T)\, e^{-i\lambda(f)} \qquad \text{(A1.15)}$$

If T is not infinite, this will contain fluctuations arising from frequencies in the range $f \lesssim 1/T$. These are the residual noise fluctuations that remain after *finite* time averaging. As T is increased, i.e., as the integrating time constant is increased, the noise spectrum becomes narrower until, in the limit $T \to \infty$, when $\mathscr{L}(f, T)\, e^{-i\lambda(f)}$ is replaced in Eq. (A1.15) by $(1/2T)\delta(f)$, the steady state is reached—cf. Eqs. (A1.8) and (A1.9). For the spectral energy distribution we have

$$\rho(\nu) = \langle \rho(\nu) \rangle_{T\to\infty} = \lim_{T\to\infty} \frac{\varepsilon_0 \tilde{\mathbf{E}}(\nu) \cdot \tilde{\mathbf{E}}^*(\nu)}{T} \tag{A1.16}$$

The energy density is

$$u = \int_0^{\infty} d\nu\, \rho(\nu) = \frac{1}{2} \int_{-\infty}^{+\infty} d\nu\, \rho(\nu) \tag{A1.17}$$

The spectral intensity distribution is

$$s(\nu) = c\rho(\nu) = \lim_{T\to\infty} \frac{\tilde{\mathbf{E}}(\nu) \cdot \tilde{\mathbf{E}}^*(\nu)}{\mu_0 c T} \tag{A1.18}$$

The total intensity is

$$I = cu = \int_0^{\infty} d\nu\, s(\nu) = \int_0^{\infty} d\nu\, c\rho(\nu) \tag{A1.19}$$

A1.5. THE MONOCHROMATIC FIELD

For a monochromatic field of frequency ν_1 we have

$$F(t) \Rightarrow \mathbf{E}(t) = \frac{E_0}{\sqrt{2}} (\boldsymbol{\pi}^+ e^{-i2\pi\nu_1 t} + \boldsymbol{\pi}^- e^{+i2\pi\nu_1 t})$$

The Fourier transform of $\mathbf{E}(t)$ is

$$\tilde{\mathbf{E}}(\nu) = \frac{E_0}{\sqrt{2}} [\boldsymbol{\pi}^+ \delta(\nu - \nu_1) + \boldsymbol{\pi}^- \delta(\nu + \nu_1)]$$

Placed in Eq. (A1.15), this yields (after some tedious algebra)

$$\begin{aligned} \langle \rho(\nu, t) \rangle_T = \varepsilon_0 E_0^2 \delta(\nu - \nu_1) \{ &\mathscr{L}(-) \cos[2\pi(\nu - \nu_1)t - \lambda(-)] \\ &+ \cos(2\alpha) \mathscr{L}(+) \cos[2\pi(\nu + \nu_1)t - \lambda(+)] \} \\ &+ \text{conjugate terms with } \nu \to -\nu \end{aligned} \tag{A1.20}$$

This distribution is "spiked" at $\nu = \nu_1$. The notations $\mathscr{L}(\pm)$ and $\lambda(\pm)$ have been used for the factors $\mathscr{L}(f, T)$ and $\lambda(f)$ with $f = \nu \pm \nu_1$. The angle α which occurs in the factor $\cos(2\alpha)$ is the polarization defining parameter introduced in Eqs. (3.14). If one uses the full expression with the conjugate term, one must use the expression for the energy density in Eq. (A1.17) with the integration over positive and negative frequencies. Alternatively, the first half of Eq. (A1.20) can be used with integration over positive frequencies only. Either process yields the energy density:

$$\begin{aligned}\langle u(t)\rangle_T &= \varepsilon_0 E_0^2\{1 + \cos(2\alpha)\mathscr{L}(2\nu_1)\cos[4\pi\nu_1 t - \lambda(2\nu_1)]\}\\ &= \varepsilon_0 E_0^2\{1 + \cos(2\alpha)(1 + 16\pi^2\nu_1^2 T^2)^{-1/2}\\ &\quad \times \cos[4\pi\nu_1 t - \arctan(4\pi\nu_1 T)]\}\end{aligned} \tag{A1.21}$$

where, in the last line, the specific Lorentzian forms for $\mathscr{L}(f, T)$ and $\lambda(f)$, given in Eq. (A1.12), have been used.

The residual oscillation, at frequency $2\nu_1$, is that which was noted in Eq. (5.3) for the intensity density of a monochromatic wave; it disappears as the averaging time $T \to \infty$. Note that it disappears also for $\alpha = 45°$, the value for the polarization defining parameter which gives circularly polarized radiation, as in Eq. (5.4). In these circumstances Eq. (A1.20) tends to

$$u = \varepsilon_0 E_0^2 = \frac{E_0^2}{\mu_0 c^2}$$

the standard result for the energy density of monochromatic radiation. The intensity of the radiation is given by

$$I = cu = \frac{E_0}{\mu_0 c}$$

A1.6. THE MODULATED FIELD ARISING FROM THE COHERENT SUPERPOSITION OF TWO FREQUENCIES

$$E(t) = \sqrt{2}E_1\cos(2\pi\nu_1 t + \Theta_1) + \sqrt{2}E_2\cos(2\pi\nu_2 t + \Theta_2) \tag{A1.22}$$

The algebra is lengthy and even more tedious. The expression for $\langle\rho(\nu, t)\rangle_T$ involves "spikes" at frequencies $\pm\nu_1$ and $\pm\nu_2$. After integrating over ν to obtain the energy density in the field, one obtains

$$\begin{aligned}\langle u(t)\rangle_T &= \varepsilon_0 E_1^2 + \varepsilon_0 E_2^2 \\ &\quad + 2\varepsilon_0 E_1 E_2 \mathscr{L}(-)\cos[2\pi(\nu_1 - \nu_2)t + (\Theta_1 - \Theta_2) - \lambda(-)] \\ &\quad + \varepsilon_0 E_1^2 \mathscr{L}(2\nu_1)\cos[4\pi\nu_1 t + 2\Theta_1 - \lambda(2\nu_1)] \\ &\quad + \varepsilon_0 E_2^2 \mathscr{L}(2\nu_2)\cos[4\pi\nu_2 t + 2\Theta_2 - \lambda(2\nu_2)] \\ &\quad + 2\varepsilon_0 E_1 E_2 \mathscr{L}(+)\cos[2\pi(\nu_1 + \nu_2)t + (\Theta_1 + \Theta_2) - \lambda(+)] \end{aligned} \tag{A1.23}$$

The notations $\mathscr{L}(\pm)$ and $\lambda(\pm)$ are short-hand for arguments $(\nu_1 \pm \nu_2)$. The last three terms are the residual oscillations at high frequencies, $2\nu_1, 2\nu_2$, and $\nu_1 + \nu_2$, which remain after averaging over an interval T; they become negligible for $\nu_1 T \gg 1$, $\nu_2 T \gg 1$. The remaining terms are the standard result for the superposition of two frequencies, i.e., the sum of the intensities of the two component waves *plus a beat or interference term.*

The result of Eq. (A1.23) can be directly generalized to the superposition of more frequencies. For example, a signal of carrier frequency ν that is amplitude modulated at frequency f can be written as a superposition of three frequencies. If m is the *amplitude of modulation,* we have

$$\begin{aligned}E(t) &= \sqrt{2}\,E_0[1 + m\cos(2\pi f t)]\cos(2\pi\nu t) \\ &= \sqrt{2}\,E_0\cos(2\pi\nu t) + \frac{mE_0}{\sqrt{2}}\cos 2\pi(\nu + f)t \\ &\quad + \frac{mE_0}{\sqrt{2}}\cos 2\pi(\nu - f)t \end{aligned} \tag{A1.24}$$

The generalization for the energy density becomes

$$\begin{aligned}\langle u(t)\rangle_T = \varepsilon_0 E_0^2 \bigg\{ 1 &+ \frac{m^2}{2} + 2m\mathscr{L}(f, T)\cos[2\pi f t - \lambda(f)] \\ &+ \frac{m^2}{2}\mathscr{L}(2f, T)\cos[4\pi f t - \lambda(2f)] \\ &+ \text{high-frequency terms} \bigg\} \end{aligned} \tag{A1.25}$$

The high-frequency terms can be neglected since they contain factors like $\mathscr{L}(2\nu, T)$ etc., which tend to zero if T is large enough such that ($\nu T \gg 1$). However, let us presume that $fT \ll 1$ so that $\mathscr{L}(f, T)$ and $\mathscr{L}(2f, T)$ are close to unity and the modulation terms remain. Then we have

$$u = \varepsilon_0 E_0^2 \left[1 + \frac{m^2}{2} + 2m\cos(2\pi f t) + \frac{m^2}{2}\cos(4\pi f t) \right]$$

This is the same result as would be obtained from $u = \varepsilon_0 E^2(t)$, by using Eq. (A1.22) for $E(t)$, and replacing $\cos^2(2\pi\nu t)$ by its average value, 1/2.

We have demonstrated that the spectral distribution for a field $X(t) = CF(t)F(t)$, averaged over an integrating period T, is given in general by Eq. (A1.14); the quantities $\mathscr{L}(f, T)$ and $\lambda(f)$, given here for exponential averaging and Lorentz resonance, will have different forms depending on how the averaging is carried out.

APPENDIX 2

Ensemble Averaging

In some cases when we are dealing with a stochastic field, i.e., with a superposition of statistically independent monochromatic components, it is necessary to use a more sophisticated form of averaging—called ensemble averaging. Averaging is performed over the random phases in the frequency components of the field. In many circumstances this is equivalent to time averaging.

We begin by considering a more general bilinear product than that given in Eq. (A1.1):

$$\begin{aligned} X(\mathbf{x}_1, t_1; \mathbf{x}_2, t_2) &= CF(\mathbf{x}_1, t_1)F(\mathbf{x}_2, t_2) \\ X(\mathbf{x}, \xi; t, \tau) &= CF(\mathbf{x}, t)F(\mathbf{x} - \xi, t - \tau) \end{aligned} \tag{A2.1}$$

The spatial cross correlation is of importance in spatial interference situations, but we shall confine our interest to correlations in time only. Furthermore, we shall be interested primarily in the properties of the electromagnetic field, so we shall immediately specialize our discussion to the energy density represented by the product

$$u(\mathbf{x}; t, \tau) = \varepsilon_0 \mathbf{E}(\mathbf{x}, t) \cdot \mathbf{E}(\mathbf{x}, t - \tau) \tag{A2.2}$$

This represents the correlation that exists between the fields at the same point in space $\mathbf{x}$, but at two different times t and $t - \tau$. In general the electric field is a superposition of monochromatic components. Each component has a frequency ν; each component has a direction of propagation specified by the *unit propagation vector* $\mathbf{k}^0 = (\sin\theta\cos\phi, \sin\theta\sin\phi, \cos\theta)$, where θ and ϕ are the polar and azimuthal angles of propagation; each component has a polarization direction specified by the *unit polarization vector* $\boldsymbol{\pi}^0$; and each has a statistically random phase specified by the *phase angle* φ. Each monochromatic component has an amplitude factor $g(\nu\mathbf{k}^0\lambda)$, which is, in general, a function of the frequency ν, the direction of propagation $\mathbf{k}^0$, and the polarization parameter λ. The quantities $\boldsymbol{\pi}^0$ and φ are also, in general, functions of these parameters. Thus, strictly, we have

$$\boldsymbol{\pi}^0 = \boldsymbol{\pi}^0(\nu \mathbf{k}^0 \lambda)$$

$$\varphi = \varphi(\nu \mathbf{k}^0 \lambda)$$

but we shall simplify the writing of equations by omitting specific reference to these functional dependences in the following. (Note the difference: ϕ is the azimuthal coordinate angle; φ is the phase angle.)

The general expansion for the electric field is

$$\mathbf{E}(\mathbf{x}, t) = \sum_{\lambda} \int_{-1}^{+1} d(\cos\theta) \int_0^{2\pi} d\phi \int_0^{\infty} d\nu\, \nu^2 g(\nu \mathbf{k}^0 \lambda)\, \boldsymbol{\pi}^0 \times \cos(2\pi\nu t - 2\pi\nu \mathbf{k}^0 \cdot \mathbf{x}/c - \varphi) \tag{A2.3}$$

- The parameter λ allows for two independent states of polarization for each component wave, with the restriction $\boldsymbol{\pi}^0(\nu \mathbf{k}^0 \lambda) \cdot \boldsymbol{\pi}^0(\nu \mathbf{k}^0 \lambda_1) = \delta(\lambda\lambda_1)$.
- The polarization vector and the propagation vector are orthogonal, $\boldsymbol{\pi}^0(\nu \mathbf{k}^0 \lambda) \cdot \mathbf{k}^0 = 0$.
- The quantity $\sin\theta\, d\theta\, d\phi\, \nu^2\, d\nu = -d(\cos\theta)\, d\phi\, \nu^2\, d\nu$ is an element of frequency-propagation space (sometimes referred to as phase space).

After substitution of this expansion into Eq. (A2.2) we are interested only in an average over the unobservable phases. This removes the rapid variations over space and time, and leaves the result in the form of integrations over distributions of frequencies and directions—this is the ensemble average. For the case we are considering it is called the mutual coherence function of the field:

$$\langle u(\tau)\rangle_\varphi = \varepsilon_0 \langle \mathbf{E}(\mathbf{x}, t) \cdot \mathbf{E}(\mathbf{x}, t - \tau)\rangle_\varphi \tag{A2.4}$$

The subscript indicates an averaging over the random phases. In the most general circumstances $\langle u\rangle_\varphi$ may still be a slowly varying function of space (nonhomogeneous field), and a slowly varying function of time (modulated field). We shall consider only the homogeneous, unmodulated case.

Associated with each component of the electric field, there is a magnetic field. The expression for the magnetic field $\mathbf{B}(\mathbf{x}, t)$ is the same as Eq. (A2.3) but with the polarization vector $\boldsymbol{\pi}^0$ replaced by $(\mathbf{k}^0 \times \boldsymbol{\pi}^0)/c$.

It is useful to write Eq. (A2.3) using complex notation:

$$\mathbf{E}(\mathbf{x}, t) = \sum_{\lambda} \int_{-1}^{+1} d(\cos\theta) \int_{0}^{2\pi} d\phi \int_{0}^{\infty} d\nu\, \nu^2 g(\nu\mathbf{k}^0\lambda)\, \boldsymbol{\pi}^0$$

$$\times \tfrac{1}{2}[e^{-i2\pi\nu t}\, e^{i2\pi\nu\mathbf{k}^0\cdot\mathbf{x}/c}\, e^{i\varphi} + \text{c.c.}]$$

$$= \mathbf{E}^{+}(\mathbf{x}, t) + \mathbf{E}^{-}(\mathbf{x}, t) \qquad \text{(A2.5)}$$

where $\mathbf{E}^{+}$ and its complex conjugate $\mathbf{E}^{-}$ are analytic expressions for the field.

Now form the mutual coherence function Eq. (A2.4). After recognizing that we are dealing with a homogeneous field [which is equivalent to placing $\mathbf{x} = 0$ in Eq. (A2.5) because one point is as good as any other], and averaging to zero any rapidly varying temporal terms (with frequencies of the order 2ν), we obtain

$$\langle u(\tau)\rangle_{\varphi} = \langle \varepsilon_0 \mathbf{E}^{+}(t)\cdot\mathbf{E}^{-}(t-\tau) + \varepsilon_0 \mathbf{E}^{-}(t)\cdot\mathbf{E}^{+}(t-\tau)\rangle_{\varphi}$$

$$= \langle u(\tau)\rangle_{\varphi}^{+} + \langle u(\tau)\rangle_{\varphi}^{-} \qquad \text{(A2.6)}$$

When expressions for $\mathbf{E}^{+}$ and $\mathbf{E}^{-}$ from Eq. (A2.5) are substituted in this, the ensemble averaging bracket survives in the factor:

$$\langle e^{i\varphi(\nu\mathbf{k}^0\lambda)}\, e^{-i\varphi(\nu_1\mathbf{k}_1^0\lambda_1)}\rangle_{\varphi} = \langle e^{i\varphi}\, e^{-i\varphi_1}\rangle_{\varphi}$$

Since the phase is random, this involves the product of δ functions:

$$\langle e^{i\varphi}\, e^{-i\varphi_1}\rangle_{\varphi} = \frac{4\pi}{3}\nu\delta(\nu-\nu_1)\delta(\cos\theta - \cos\theta_1)\delta(\phi-\phi_1)\delta(\lambda\lambda_1) \qquad \text{(A2.7)}$$

The normalization has been chosen so that, firstly, the quantity is dimensionless, and secondly, the result of the integration

$$\sum_{\lambda} \int_{-1}^{+1} d(\cos\theta_1) \int_{0}^{2\pi} d\phi_1 \int_{0}^{\nu} d\nu_1\, \nu_1^2 \langle e^{i\varphi}\, e^{-i\varphi_1}\rangle = \frac{4\pi}{3}\nu^3$$

being the volume of frequency-propagation space within the value ν. Of course, the matter of normalization is, to some extent, irrelevant at this stage because the magnitude of $g(\nu\mathbf{k}^0\lambda)$ has not yet been specified.

Place now the expansion of $\mathbf{E}^+$ and $\mathbf{E}^-$ of Eq. (A2.5) into Eq. (A2.6) and use Eq. (A2.7):

$$\langle u(\tau)\rangle_\varphi^+ = \frac{\pi\varepsilon_0}{3}\sum_\lambda \int_{-1}^{+1} d(\cos\theta) \int_0^{2\pi} d\phi \int_0^\infty d\nu\, \nu^5 g^2(\nu\mathbf{k}^0\lambda)\, e^{-i2\pi\nu\tau} \qquad \text{(A2.8)}$$

At this stage the expression for the mutual coherence function is still fairly general since the "amplitude" $g(\nu\mathbf{k}^0\lambda)$ may depend on $\mathbf{k}^0$ and λ; therefore it can be appropriate to anisotropic and polarized fields. We shall, however, specialize the result to the case of the isotropic, unpolarized field. In this case $g(\nu\mathbf{k}^0\lambda)$ becomes a function of ν only; the integration over angles gives a factor of 4π steradians, and the summation gives a factor of 2 for two independent states of polarization.

Therefore, for a *homogeneous, unmodulated, isotropic, unpolarized* field, we have

$$\langle u(\tau)\rangle_\varphi^+ = \int_0^\infty d\nu \frac{8\pi^2\varepsilon_0}{3}\nu^5 g^2(\nu)\, e^{-i2\pi\nu\tau}$$

$$= \int_0^\infty d\nu\, \tfrac{1}{2}\rho(\nu)\, e^{-i2\pi\nu\tau} \qquad \text{(A2.9)}$$

The *real* mutual coherence function is

$$\langle u(\tau)\rangle_\varphi = \langle u(\tau)\rangle_\varphi^+ + \langle u(\tau)\rangle_\varphi^-$$

$$= \int_0^\infty d\nu\, \rho(\nu)\cos(2\pi\nu\tau) \qquad \text{(A2.10)}$$

For $\tau = 0$, this reduces to the *energy density* of the field:

$$\langle u(0)\rangle = \int_0^\infty d\nu\, \rho(\nu)$$

which shows that the quantity $\rho(\nu)$, introduced in Eq. (A2.9) is indeed the spectral energy density, the energy of the field per unit volume per unit frequency interval. The spectral energy density for this isotropic case is, from Eq. (A2.9),

$$\rho(\nu) = \frac{16\pi^2\varepsilon_0\nu^5}{3} g^2(\nu) = \left(\frac{\mu_0}{4\pi}\right)^{-1} \frac{4\pi\nu^5}{3c^2} g^2(\nu) \qquad \text{(A2.11)}$$

This equation relates the spectral energy density $\rho(\nu)$ to the "amplitude" distribution function $g(\nu)$. To generate a field of a certain required property, it is necessary to specify the latter.

For example, the spectral energy density of *black-body radiation* is generated by the choice

$$g^2(\nu) = \left(\frac{\mu_0}{4\pi}\right)\frac{6h}{c\nu^2}\frac{1}{e^{h\nu/kT} - 1}$$

giving

$$\rho_{BB}(\nu) = \frac{8\pi\nu^2}{c^3}\left(\frac{h\nu}{e^{h\nu/kT} - 1}\right) \tag{A2.12}$$

The first factor on the right-hand side is the number of degrees of freedom (modes) per unit volume for standing waves in a large enclosure per unit frequency interval Eq. (A7.7). The second factor is the average energy per mode for black-body radiation.

In Appendix 3 it is shown that when $g(\nu)$ is inversely proportional to the frequency ν, an isotropic radiation field is generated which has the spectral requirements for a *zero-point* vacuum field. In particular, with

$$g^2(\nu) = \frac{\mu_0}{4\pi}\frac{6h}{c\nu^2}$$

$$\rho^+_{ZP}(\nu) = \frac{1}{2}\frac{8\pi\nu^2}{c^3}h\nu \tag{A2.13}$$

The factor of $\frac{1}{2}$ in this expression is due to the fact, discussed in Appendix 3, that the zero-point field is a "half" field only, in that it causes downward transitions (spontaneous decay) but not upward transitions (spontaneous absorption). Thus the mutual coherence function consists only of the $\langle u\rangle^+$ term as in Eq. (A2.9) and not the real quantity as in Eq. (A2.10). The spectral density is therefore: $\frac{1}{2}\times$ (the number of degrees of freedom per unit volume per unit frequency interval) $\times$ (the energy per mode).

APPENDIX 3

THE ZERO-POINT VACUUM FIELD

It is believed that an electromagnetic field exists in vacuum even when the temperature of the surrounding material is reduced toward zero so that the thermal field tends to zero. The existence of such a zero-point field has been confirmed experimentally by the Casimir experiment[1] (the measurement of the attractive force between two parallel plates in an evacuated, near-zero-temperature enclosure). That force is found to be proportional to the inverse fourth power of the distance apart of the plates; it has been shown that such a result can only be produced by a zero-point field whose spectral energy density has a frequency dependence $\rho(\nu) = k\nu^3$. The value of the constant in the spectral density equation cannot be calculated theoretically, but the experimental measurement shows that, after factoring out some numerical constants, h (Planck's constant) is involved and the measured value of k is in agreement with the expression $4\pi h/3c^2$. Thus Planck's constant, a hallmark of quantum physics, appears in a purely classical context. The situation is not unlike that which emerged in Section 6, where a classical theory of the Compton effect led to the emergence of a constant which, experimentally, had the value of Planck's constant. (For a very readable account, see Boyer[2]; see also Boyer.[3])

Some characteristics of zero-point radiation can be deduced simply from the fact that it exists in vacuum. The nature of vacuum demands that no special place or direction in it is defined. Hence the radiation, like thermal radiation, must be homogeneous and isotropic. Furthermore, the vacuum should not define any special velocity; it should look the same to any two observers irrespective of their relative velocity. Thus the zero-point radiation spectrum must be Lorentz invariant. This fact demonstrates at once that the zero-point field cannot have the character of black-body radiation, for example, nor can it have any peak in its spectral distribution. Such a feature would be shifted to different parts of the spectrum for different observers and could therefore be used to distinguish between them. And what may have been isotropic for one observer would not be for another.

It turns out that the only field that obeys Lorentz invariance is one whose spectral energy density involves a factor ν^3. If one changes one's velocity in such a field, the Doppler shift in the frequencies is completely counteracted by the relativistic changes of intensity so that an identical spectrum is reproduced. This rather surprising result will be demonstrated in what follows.

The general expression for the electric field has been given in Eq. (A2.3) and Eq. (A2.5). We shall show that the choice

$$g(\nu \mathbf{k}^0 \lambda) = g(\nu) = b/\nu \tag{A3.1}$$

b being a Lorentz-invariant physical quantity, leads to a field that is Lorentz invariant.

We therefore place Eq. (A3.1) into Eq. (A2.5) and obtain an expression for the zero-point electric field (we also place $x = 0$ because of its homogeneous nature):

$$\begin{aligned} \mathbf{E}_{ZP}(t) &= \frac{b}{2}\sum_{\lambda} \int_{-1}^{+1} d(\cos\theta) \int_0^{2\pi} d\phi \\ &\quad \times \int_0^{\infty} d\nu\, \nu\, \boldsymbol{\pi}^0(\nu \mathbf{k}^0 \lambda)[e^{-i2\pi\nu t}\, e^{+i\varphi(\nu \mathbf{k}^0 \lambda)} + \text{c.c.}] \\ &= \mathbf{E}^+_{ZP}(t) + \mathbf{E}^-_{ZP}(t) \end{aligned} \tag{A3.2}$$

This is associated with a magnetic field

$$\begin{aligned} \mathbf{B}_z(t) &= \frac{b}{2c}\sum_{\lambda} \int_{-1}^{+1} d(\cos\theta) \int_0^{2\pi} d\phi \\ &\quad \times \int_0^{\infty} d\nu\, (\mathbf{k}^0 \times \boldsymbol{\pi}^0)[e^{-i2\pi\nu t}\, e^{+i\varphi(\nu \mathbf{k}^0 \lambda)} + \text{c.c.}] \end{aligned} \tag{A3.3}$$

The energy density of the field is

$$\begin{aligned} \langle u_{ZP} \rangle &= \langle u_{ZP} \rangle^+ + \langle u_{ZP} \rangle^- \\ &= \varepsilon_0 \langle \mathbf{E}^+_{ZP} \cdot \mathbf{E}^-_{ZP} \rangle + \varepsilon_0 \langle \mathbf{E}^-_{ZP} \cdot \mathbf{E}^+_{ZP} \rangle \end{aligned} \tag{A3.4}$$

The two terms on the right-hand side are equal; we shall here treat one only, reserving for later discussion how to establish a value for the energy density:

$$\langle u_{\text{ZP}}\rangle^{+} = \frac{\varepsilon_0 b^2}{4}\sum_{\lambda}\int_{-1}^{+1} d(\cos\theta)\int_0^{2\pi} d\phi\int_0^{\infty} d\nu\, \nu\, \boldsymbol{\pi}^0 \cdot \sum_{\lambda_1}\int_{-1}^{+1} d(\cos\theta_1)$$

$$\times \int_0^{\infty} d\phi_1 \int_0^{\infty} d\nu_1\, \nu_1 \boldsymbol{\pi}_1^0\, e^{-i2\pi(\nu-\nu_1)t}\langle e^{i\varphi(\nu\mathbf{k}^0\lambda)}\, e^{-i\varphi(\nu_1\mathbf{k}_1^0\lambda_1)}\rangle_{\varphi}$$

Using the result of Eq. (A2.7) and carrying out the integrations, we obtain

$$\langle u_{\text{ZP}}\rangle^{+} = \frac{\pi\varepsilon_0 b^2}{3}\sum_{\lambda}\int_{-1}^{+1} d(\cos\theta)\int_0^{2\pi} d\phi\int_0^{\infty} d\nu\, \nu^3$$

$$= \int_0^{\infty} d\nu\, \rho_{\text{ZP}}^{+}(\nu) \tag{A3.5}$$

with

$$\rho_{\text{ZP}}^{+} = \frac{8\pi^2\varepsilon_0 b^2 \nu^3}{3} = \left(\frac{\mu_0}{4\pi}\right)^{-1}\frac{2\pi b^2}{3c^2}\nu^3 \tag{A3.6}$$

Thus, our chosen equation has produced a ν^3 dependence; the value of b still has to be specified. We note that the energy density is only half that of Eq. (A2.11), reflecting the fact that we are considering only one half of a real field, Eq. (A3.5).

We are now ready to show that such a field is Lorentz invariant. Make a Lorentz transformation of the fields from the point of view of a new frame of reference (primed quantities) traveling with velocity $\mathbf{V} = \beta c \mathbf{z}^0$ relative to the above (the direction of the z axis is purely arbitrary in an isotropic situation, of course).

$$\mathbf{E}'_{\text{ZP}} = \frac{b}{2}\sum_{\lambda}\int_{-1}^{+1} d(\cos\theta')\int_0^{2\pi} d\phi'\int_0^{\infty} d\nu'\, \nu'\boldsymbol{\pi}^0(\nu'\mathbf{k}^{0\prime}\lambda)[e^{-i2\pi\nu' t'}\, e^{i\varphi'} + \text{c.c.}]$$

$$= \mathbf{x}^0 E'_x + \mathbf{y}^0 E'_y + \mathbf{z}^0 E'_z$$

$$= \mathbf{x}^0\gamma(E_x - \beta c B_y) + \mathbf{y}^0\gamma(E_y + \beta c B_x) + \mathbf{z}^0 E_z$$

$$= \frac{b}{2}\sum_{\lambda}\int_{-1}^{+1} d(\cos\theta)\int_0^{2\pi} d\phi\int_0^{\infty} d\nu\, \nu\, \boldsymbol{\pi}^*[e^{-i2\pi\nu t}\, e^{i\varphi} + \text{c.c.}] \tag{A3.7}$$

The first line is the expansion of $\mathbf{E}'_{\text{ZP}}$ in terms of primed coordinates, etc. The second line expresses this field in terms of its Cartesian components. The step between line 2 and line 3 is just the relativistic transformation for these into components of the unprimed frame; the step between line 3 and line 4 involves the substitution of Eqs. (A3.2) and (A3.3) for the components of the fields. The new propagation vector $\boldsymbol{\pi}^*$ can be expressed in terms of the x-, y-, and z-components of $\boldsymbol{\pi}^0$ and $\mathbf{k}^0$ (note: $\boldsymbol{\pi}^*$ is not a unit vector):

$$\boldsymbol{\pi}^* = \mathbf{x}^0\gamma[\pi_x^0 - \beta(\mathbf{k}^0 \times \boldsymbol{\pi}^0)_y] + \mathbf{y}^0\gamma[\pi_y^0 + \beta(\mathbf{k}^0 \times \boldsymbol{\pi}^0)_x] + \mathbf{z}^0\pi_z^0 \quad \text{(A3.8)}$$

with $\gamma = (1 - \beta^2)^{-1/2}$.

The average energy density of the field in the primed frame is established from

$$\langle u'_{ZP}\rangle^+ = \varepsilon_0\langle \mathbf{E}'^+_{ZP}(t') \cdot \mathbf{E}'^-_{ZP}(t')\rangle \quad \text{(A3.9)}$$

The result is the same as Eq. (A3.5) with $\boldsymbol{\pi}^0 \cdot \boldsymbol{\pi}^0 = 1$ replaced by

$$\begin{aligned}\boldsymbol{\pi}^* \cdot \boldsymbol{\pi}^* &= \gamma^2[\pi_x^0 - \beta(\mathbf{k}^0 \times \boldsymbol{\pi}^0)_y]^2 + \gamma^2[\pi_y^0 + \beta(\mathbf{k}^0 \times \boldsymbol{\pi}^0)_x]^2 + [\pi_z^0]^2 \\ &= \gamma^2(1 - \beta\cos\theta)^2\end{aligned}$$

We thereby arrive at an expression for the average energy density in the primed frame, expressed in terms of the frequency and coordinates of the unprimed frame [cf. Eq. (A3.5)]:

$$\langle u'_{ZP}\rangle^+ = \frac{\pi\varepsilon_0 b^2}{3}\sum_\lambda \int_{-1}^{+1} d(\cos\theta) \int_0^{2\pi} d\phi \int_0^\infty d\nu\, \nu^3\gamma^2(1 - \beta\cos\theta)^2 \quad \text{(A3.10)}$$

We now transfer the right-hand side back to integrals over variables in primed space using the standard results of relativistic transformation:

$$\nu = \nu'\gamma(1 + \beta\cos\theta')$$

$$d\nu = d\nu'\,\gamma(1 + \beta\cos\theta')$$

$$\cos\theta = \frac{\cos\theta' + \beta}{1 + \beta\cos\theta'}$$

$$d(\cos\theta) = \frac{d(\cos\theta')}{\gamma^2(1 + \beta\cos\theta')^2}$$

$$\gamma^2(1 - \beta\cos\theta)^2 = \frac{1}{\gamma^2(1 + \beta\cos\theta')^2}$$

We deduce

$$\langle u'_{ZP}\rangle^+ = \int_0^\infty d\nu\, \rho^{+\prime}_{ZP}(\nu') \quad \text{(A3.11)}$$

where

$$\rho^{+\prime}_{ZP}(\nu') = \left(\frac{\mu_0}{4\pi}\right)^{-1} \frac{2\pi b^2}{3c^2} (\nu')^3 \tag{A3.12}$$

Being real, $\rho^{+\prime}_{ZP}(\nu') = \rho^{-\prime}_{ZP}(\nu')$.

We have shown that the spectral density in the primed frame, $\rho'(\nu')$, is exactly the same function of the frequency in that frame as was the case in the unprimed frame, Eq. (A3.6). Therefore an isotropic spectral density as expressed by Eq. (A3.6) or Eq. (A3.12) is Lorentz invariant. The value of the Lorentz invariant quantity b, which through Eq. (A3.2) defines the amplitude of the field, is not yet specified.

We have established expressions for the field vector, Eq. (A3.2), and its spectral energy density, Eq. (A3.6), which have the properties of homogeneity, isotropy, and Lorentz invariance that are required for zero-point radiation in a vacuum. Such radiation should not be able to excite an atom from a lower to a higher energy state. This would imply that energy could be extracted from the vacuum—a contradiction in meaning. However, such radiation could be responsible for inducing a transition from a higher to a lower energy state of an excited atom placed in the vacuum; after the transition, the space would contain extra radiation.

In order to obtain this "one-way" property it is demonstrated in Chapter 10 that only the ordering $\mathbf{E}^+(t) \cdot \mathbf{E}^-(t)$ of the analytic field components can be allowed. Thus only the "half" spectral energy density

$$\langle u_{ZP}\rangle^+ = \varepsilon_0\langle \mathbf{E}^+(t) \cdot \mathbf{E}^-(t)\rangle$$

contributes to the zero-point vacuum field, and not its complex conjugate $\langle u_{ZP}\rangle^-$. We expect then that the spectral energy density would be obtained by multiplying *half* the normal mode density $\frac{1}{2}(8\pi\nu^2/c^3)$ by the energy per mode $h\nu$ [see Eq. (A7.7)]:

$$\rho^+_{ZP}(\nu) = \frac{4\pi h\nu^3}{c^3} \tag{A3.13}$$

Comparing this with Eq. (A3.6), we obtain

$$b^2 = \left(\frac{\mu_0}{4\pi}\right) \frac{6h}{c} \tag{A3.14}$$

and

$$g^2(\nu) = \frac{b^2}{\nu^2} = \left(\frac{\mu_0}{4\pi}\right) \frac{6h}{c\nu^2}$$

This is exactly the result used in Chapter 10 to produce Eq. (10.30). [Alternatively, we could insist on the ordering E^+E^-, and write Eq. (A3.4) as $\langle u \rangle = 2\langle u \rangle^+$; then take the normal mode density $8\pi\nu^2/c^3$, and multiply by $h\nu/2$—cf. the zero energy of the quantum harmonic oscillator.]

Equations (A3.2) and (A3.14) give a description of the zero-point vacuum field which is homogeneous and isotropic and has a spectral density given by Eq. (A3.13). The field has the half modal energy density expected for a zero-point field, it is exactly that necessary to predict the experimentally observed magnitude of the Casimir effect, and, as is shown in Chapter 10, it interacts with an excited atom to produce spontaneous decay at the same rate as is predicted by the Einstein A coefficient.

REFERENCES

1. H. B. G. CASIMIR, "The van der Waals-London Forces," *J. Chem. Phys.* (*Paris*) **46**, 407–409 (1949).
2. T. H. BOYER, "The Classical Vacuum," *Sci. Am.* **253**, 56–62 (1985).
3. T. H. BOYER, "Derivation of the Black-Body Radiation Spectrum without Quantum Assumptions," *Phys. Rev.* **182**, 1374–1383 (1969).

APPENDIX 4

AN INVARIANT FORM FOR ANGULAR MOMENTUM

JUSTIFICATION FOR EQ. (6.17)

For linear motion, the invariant equation relating the linear momentum and the total energy of a free particle is written

$$-P^2 + \frac{W^2}{c^2} = m_0^2 c^2 \qquad \text{(A4.1)}$$

or, in the notation of covariant 4-vectors, $P^\mu = (\mathbf{P}, W/c)$, $P_\mu = (-\mathbf{P}, W/c)$,

$$P^\mu P_\mu = m_0^2 c^2$$

Equation (A4.1) can be written as

$$\frac{P^2}{2m} = \frac{W^2 - m_0^2 c^4}{2W} \qquad \text{(A4.2)}$$

where $m = W/c^2$ is the relativistic mass. The right-hand side is nonzero only when the system has some form of *dynamical* energy (i.e., energy other than rest-mass energy). The left-hand side, $P^2/2m$, tends to the *kinetic* energy for a free particle in the low-velocity (nonrelativistic) limit. For a free particle, the total dynamical energy *is* its kinetic energy.

An equivalent expression for a rotating particle would be $L^2/2I$, where L is the *angular* momentum and I is the *moment of inertia.* This also tends to the kinetic energy of an orbiting particle in the nonrelativistic limit. However, it must be remembered that such a particle (orbiting) is not free, but bound by forces to the center of rotation; therefore it has *potential* energy as well as kinetic energy. The total dynamical energy will be greater than $L^2/2I$. In the present case the potential energy is proportional to the square of the displacement from the center of rotation (this can readily be established by the integration of Eq. (6.8) from the center to the final radius). The virial theorem then gives the result that the potential energy equals the kinetic energy; the total dynamical energy is therefore *twice* the kinetic energy. If the left-hand side of the new equation we are trying to construct is to represent the total dynamical energy of an orbiting mass, it must be L^2/I rather than $L^2/2I$. Writing $I = L/\omega$, we obtain

$$L = \frac{W^2 - m_0^2c^4}{2\omega W} \tag{A4.3}$$

An invariant form of this, involving covariant 4-vectors, can be written

$$L = \frac{P^\mu P_\mu - m_0^2c^2}{2k^\mu P_\mu} \tag{A4.4}$$

where P^μ is the 4-momentum of the particle, and $k^\mu = (\mathbf{k}, \omega/c)$ is the 4-vector of electromagnetic wave propagation with $|\mathbf{k}| = \omega/c$. This electromagnetic wave provides the driving frequency for the rotating particle. Note that, because the particle is not free, the numerator is not zero.

Expanding the inner products, we obtain

$$L = \frac{W^2 - c^2P^2 - m_0^2c^4}{2\omega(W - cP)} \qquad \text{(Q.E.D.)} \tag{A4.5}$$

This is the formula that was quoted in Eq. (6.17) and was used to establish the quantum rule for the change of angular momentum when radiation interacts with matter. $P = |\mathbf{P}|$ is the linear momentum of the particle; W is the total energy of the particle with contributions from its rest mass, its linear motion, and its rotational motion; ω is the angular frequency of the passing radiation which maintains the particle in this stationary state. The equation can be evaluated in any frame of reference, the ZM frame discussed in Chapter 6 being the simplest because then, $P = 0$. But Eq. (A4.5), because of its Lorentz invariance, yields the same value in all frames.

APPENDIX 5

FUNCTIONS

A5.1. THE DIRAC DELTA FUNCTION

The δ function is a powerful mathematical tool in physics. It is defined by the following:

$$\delta(f) = 0 \qquad \text{if } f \neq 0 \tag{A5.1}$$

$$\int_{-\infty}^{+\infty} \delta(f)\, df = 1 \tag{A5.2}$$

The δ function has physical dimensions of $[f]^{-1}$. Evidently $\delta(f)$ is not an ordinary mathematical function, being zero everywhere except at one point; at that point its value must be infinite in order to give the finite integral. It is more appropriate to regard $\delta(f)$ as a quantity with a symbolic meaning. It can be thought of as the extreme (limiting) case of a class of functions, $F(f, T)$, which are peaked at $f = 0$, which have a height that is proportional to the parameter T, which have a width that is inversely proportional to T, and which are normalized so that, whatever the value of T,

$$\int_{-\infty}^{+\infty} F(f, T)\, df = 1 \tag{A5.3}$$

Such a function could be, for example, the *Gaussian function.*

We start by defining a Gaussian function of *time*:

$$G(t, T) = \frac{1}{2T} e^{-\pi t^2/4T^2} \tag{A5.4}$$

The Gaussian function has a mean value over all time of unity:

$$\int_{-\infty}^{\infty} G(t, T)\, dt = 1 \tag{A5.5}$$

Associated with this Gaussian is its Fourier transform:

$$\tilde{G}(f, T) = \int_{-\infty}^{+\infty} dt\, G(t, T)\, e^{i2\pi ft}$$

$$= e^{-4\pi f^2 T^2} = \mathscr{G}(f, T) \tag{A5.6}$$

It may be noted that the exponents in these Gaussian functions are written containing a factor π, rather than the π^2 that may be expected; when working with Gaussian functions, factors of $\sqrt{\pi}$ associated with the width parameter T tend to float about. This method of writing gives the same normalizations and average values as the other functions to be introduced.

The quantity $2T\mathscr{G}(f, T) = 2Te^{-4\pi f^2 T^2}$ has the following properties:

- It has a height of $2T$ at the peak $f = 0$.
- It has a half-width at half maximum (HWHM) equal to

$$\frac{1}{2T}\left(\frac{\ln 2}{\pi}\right)^{1/2} = \frac{0.2349\ldots}{T}$$

- And

$$\int_{-\infty}^{\infty} 2T\mathscr{G}(f, T)\, df = 1$$

Therefore, as T becomes larger, the height increases toward infinity and the width decreases toward zero, while the area under the curve remains constant equal to unity. Although the function $2T\mathscr{G}(f, T)$ as a proper mathematical function does not exist for all values of f as $T \to \infty$, it approaches the properties of the δ function in this limit. One can formally write

$$\lim_{T\to\infty} \mathscr{G}(f, T) = \frac{1}{2T}\delta(f) \tag{A5.7}$$

provided that it is realized that both sides of the equation are understood to participate in an integration over f with a large value for T ($T \gg 1/f$) and that, after integration, the limit $T \to \infty$ can be taken. In other words, under these instructions $2T\mathscr{G}(f, T)$ behaves like the δ function $\delta(f)$.

In the above, the symbols f and T imply frequency and time interval, respectively; certainly in many examples they have that interpretation. But they could represent other pairs of appropriately conjugate variables, e.g., time t and frequency interval Δ, wave number σ and spatial interval X, etc.

An important property of the δ function is its *shifting property*:

$$\int_{-\infty}^{+\infty} \varphi(\nu)\delta(\nu-\nu_0)\,d\nu = \varphi(\nu_0) \tag{A5.8}$$

where $\varphi(\nu)$ is any continuous function of the variable ν. The validity of this can be seen by replacing $\delta(\nu-\nu_0)$ by $2T\mathcal{G}(\nu-\nu_0, T)$ and examining the integral for large values of T. Evidently, when T is large, the integral depends only on the value of φ close to $\nu = \nu_0$; it may then be replaced by $\varphi(\nu_0)$ and removed from the integral. Using Eq. (A5.2), the result follows. This means that multiplying a continuous function of ν by $\delta(\nu-\nu_0)$ and integrating over *all* ν is equivalent to replacing ν by ν_0 in the argument of the function. Actually, it is not necessary for the range of integration to be from $-\infty$ to $+\infty$; because $\delta(\nu-\nu_0)$ is, according to its definition, equal to zero at every value of ν except ν_0, it is only necessary for the range of integration to contain ν_0 in its interior. The result can also be written symbolically as

$$\varphi(\nu)\delta(\nu-\nu_0) = \varphi(\nu_0)\delta(\nu-\nu_0) \tag{A5.9}$$

meaning that the two sides yield the same result when used in a subsequent integration over an appropriate range of ν.

Another representation of the δ function can be found by placing $\varphi(\nu) = e^{i2\pi\nu t}$ in Eq. (A5.8). Then, after placing $\nu_0 = 0$, we have

$$\int_{-\infty}^{+\infty} e^{i2\pi\nu t}\,\delta(\nu)\,d\nu = 1 \tag{A5.10}$$

This may be interpreted as "unity is the Fourier transform of the δ function." Then, by the reciprocal relation, "the δ function is the Fourier transform of unity":

$$\delta(\nu) = \int_{-\infty}^{+\infty} e^{-i2\pi\nu t}\,dt \tag{A5.11}$$

We have introduced the peaked function, the Gaussian $\mathcal{G}(f, T)$, to illustrate and justify the properties of the δ function. Other peaked functions can also be used, e.g., the Lorentz "resonance" function, and the Fraunhofer "diffraction" function. Since these functions are used in the text an equivalent treatment of their properties is now discussed.

A5.2. THE EXPONENTIAL AND THE LORENTZ FUNCTIONS

Exponential functions play a vital role in physics. We meet them, for example, in the decay of some property of a system from its initial value at time t_0; the exponential decay factor is $e^{-(t-t_0)/T}$. T is the mean life of the function. If the property is the stored energy of an oscillating system, $T = 1/\gamma$, where γ is the decay constant; if the property is the amplitude of the oscillating system, $T = \gamma/2$. We meet the exponential function also in the form $e^{+(t-t_0)/T}$. This represents an exponential rise or a *decaying memory* of things that happened before t_0.

When these two functions are placed back-to-back, we can define a peaked time function (place $t_0 = 0$):

$$L(t, T) = \frac{1}{2T} e^{+t/T}, \qquad t \le 0$$

$$= \frac{1}{2T} e^{-t/T}, \qquad t \ge 0 \tag{A5.12}$$

where the function has been normalized with the factor $1/2T$ so that

$$\int_{-\infty}^{+\infty} L(t, T)\, dt = 1 \tag{A5.13}$$

The Fourier transform of $L(t, T)$ is

$$\tilde{L}(f, T) = \frac{1}{1 + 4\pi^2 f^2 T^2} = \mathscr{L}^2(f, T) \tag{A5.14}$$

where $\mathscr{L}(f, T)$ will be used for the Lorentz (resonance) function $(1 + 4\pi^2 f^2 T^2)^{-1/2}$. The quantity $2T\mathscr{L}^2(f, T)$ has the following properties:

- It has a height of $2T$ at $f = 0$.
- It has a half-width at half maximum (HWHM) of

$$\Delta f = \frac{1}{2\pi T}$$

- And it is normalized according to

$$\int_{-\infty}^{+\infty} 2T\mathscr{L}^2(f, T)\, df = 1$$

Therefore, as $T \to \infty$, it behaves as the δ function and we can write

$$\lim_{T\to\infty} \mathscr{L}^2(f, T) = \frac{1}{2T}\,\delta(f) \tag{A5.15}$$

$\mathscr{L}^2(f, T)$ therefore has similar properties to $\mathscr{G}(f, T)$ as $T \to \infty$ and can be used in the same way as the Gaussian function to give an operational meaning to the δ function.

The peaked function of time defined in Eq. (A5.12) is rather artificial; more often, one meets one or the other of the half functions that go to make its definition

$$\begin{aligned} L^+(t, T) &= 0, & t < 0 \\ &= \frac{1}{T}\,e^{-t/T}, & t \geq 0 \end{aligned} \tag{A5.16}$$

representing the exponential decay of a response after it has been switched on, normalized with the factor $1/T$. The Fourier transform of $L^+(t, T)$ is

$$\begin{aligned} \tilde{L}^+(f, T) &= \frac{1}{1 - i2\pi fT} \\ &= \mathscr{L}(f, T)\,e^{+i\lambda} \end{aligned} \tag{A5.17}$$

where

$$\begin{aligned} \mathscr{L}(f, T) &= \frac{1}{(1 + 4\pi^2 f^2 T^2)^{1/2}} \\ \lambda &= \arctan(2\pi fT) \end{aligned} \tag{A5.18}$$

$\mathscr{L}(f, T)$ and λ are plotted in Fig. 32. The other "half"-function $L^-(t, T) = (1/T)\,e^{+t/T}$ for $t \leq 0$ has a Fourier transform $\tilde{L}^-(f, T) = 1/(1 + i2\pi fT) = \mathscr{L}(f, T)\,e^{-i\lambda}$. Note that

$$\tilde{L}^+(f, T) + \tilde{L}^-(f, T) = \mathscr{L}(f, T)2\cos\lambda = 2\mathscr{L}^2(f, T)$$

$\tilde{L}^+(f, T)$ and $\tilde{L}^-(f, T)$ also have properties closely related to the δ function.

$$2T\tilde{L}^+(f, T) = 2T\mathscr{L}(f, T)\,e^{i\lambda} = \frac{2T}{1 - i2\pi fT}$$

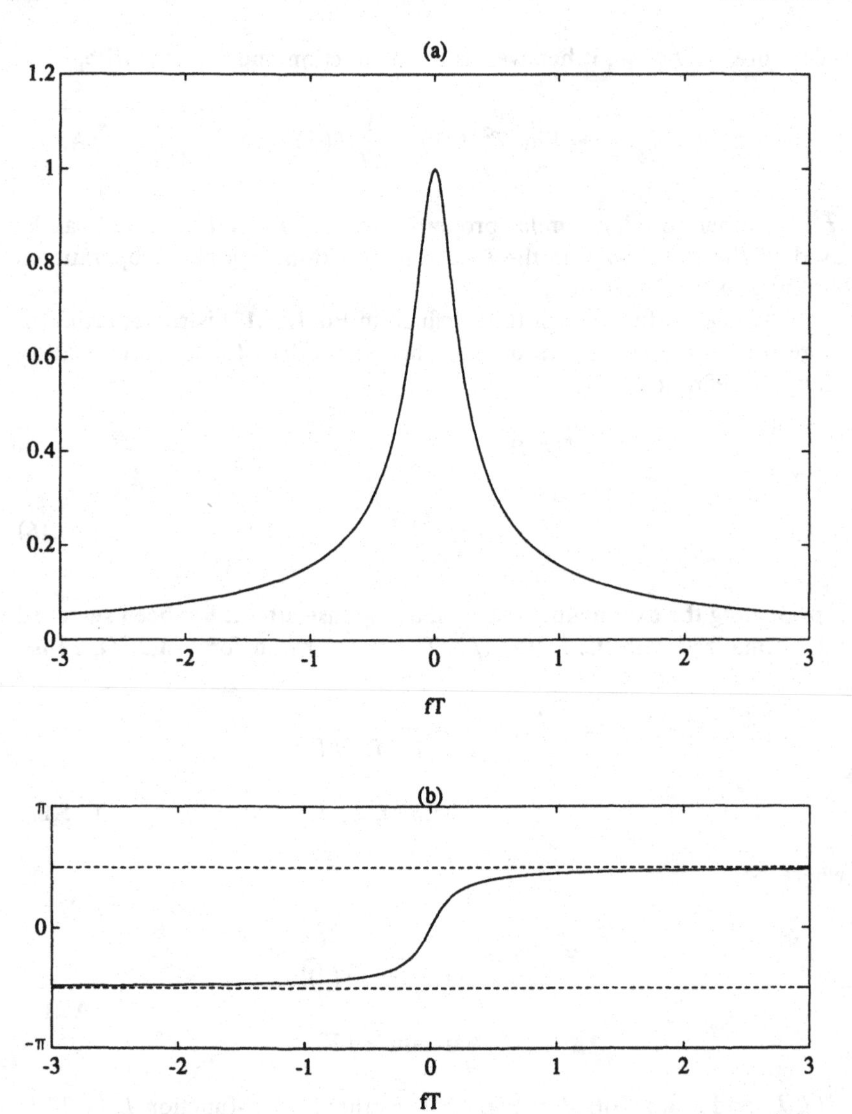

FIGURE 32. (a) The Lorentz function $\mathscr{L}(f, T)$ and (b) its phase.

has the following properties:

- It has a height of $2T$.
- It has a HWHM of $1/i2\pi T$ (inversely proportional to T, albeit imaginary).

- The integral

$$\int_{-\infty}^{+\infty} \frac{2T}{1 - i2\pi fT}\, df = 1$$

We can therefore write formally

$$\lim_{T\to\infty} \mathscr{L}(f, T)\, e^{i\lambda} = \lim_{T\to\infty} \frac{1}{(1 + 4\pi^2 f^2 T^2)^{1/2}}\, e^{\pm i\lambda}$$

$$= \frac{1}{2T}\, \delta(f) \qquad \text{(A5.19)}$$

Note that the quantity $\mathscr{L}(f, T)$ does not itself integrate over all f to a finite quantity and therefore does not tend to the δ function as $T \to \infty$; it is the presence of the extra phase factor $e^{\pm i\lambda}$ in Eq. (A5.19) that gives this property to $\mathscr{L}(f, T)\, e^{\pm i\lambda}$.

The Lorentzian may also be recognized in the analysis of the behavior of a damped oscillator. Consider the expression for the motion of a damped free system as given in Eq. (3.3), now written as

$$A(t) = 0, \qquad t < 0$$

$$= \sqrt{2}\, A_0\, e^{-\gamma t/2} \cos(2\pi\nu' t), \qquad t \geq 0$$

$$= A^+(t) + A^-(t) \qquad \text{(A5.20)}$$

where $A^+(t) = (A_0/\sqrt{2})\, e^{-\gamma t/2}\, e^{-i2\pi\nu' t}$ is the analytic "half"-field for $t \geq 0$ (the damping constant $\gamma/2$ plays the same role as the reciprocal of the time constant T). The Fourier transform is

$$\tilde{A}(\nu) = \frac{A_0}{\sqrt{2}} \left[\frac{1}{\gamma/2 - i2\pi(\nu - \nu')} + \frac{1}{\gamma/2 - i2\pi(\nu + \nu')} \right]$$

$$= \frac{\sqrt{2}\, A_0}{\gamma} \left[\tilde{L}^+\!\left(\nu - \nu', \frac{2}{\gamma}\right) + \tilde{L}^+\!\left(\nu + \nu', \frac{2}{\gamma}\right) \right] \qquad \text{(A5.21)}$$

which, using Eq. (A5.18), can be written

$$\tilde{A}(\nu) = \frac{\sqrt{2} A_0}{\gamma} [\mathscr{L}(-)\, e^{+i\lambda(-)} + \mathscr{L}(+)\, e^{+i\lambda(+)}]$$

where the notations $\mathscr{L}(\pm)$ and $\lambda(\pm)$ stand for Eq. (A5.18) with arguments containing $\nu \pm \nu'$, respectively. The superscript + on both the Fourier transforms in Eq. (A5.21) is a consequence of the function being nonzero only for positive values of t. Such Fourier expansions play a part in the analyses of resonance situations in Chapter 3 and also in the discussions of time averaging in Appendix 1.

A5.3. THE SQUARE PULSE AND THE FRAUNHOFER FUNCTION

The square pulse at time t_0 and of duration T is defined by the statement

$$\begin{aligned} p^+(t) &= 0, \qquad t < t_0 \\ &= 1, \qquad t_0 \leq t \leq t_0 + T \\ &= 0, \qquad t > t_0 + T \end{aligned}$$

It is also met with in the form, indicating a flat or truncated memory over an interval T before t_0:

$$\begin{aligned} p^-(t) &= 0, \qquad t < t_0 - T \\ &= 1, \qquad t_0 - T \leq t \leq t_0 \\ &= 0, \qquad t > t_0 \end{aligned}$$

When these two are placed back-to-back we can define a peaked time function at $t_0 = 0$:

$$\begin{aligned} F(t, T) &= 0, \qquad t < -T \\ &= 1/2T, \qquad -T < t + T \\ &= 0, \qquad t > T \end{aligned} \tag{A5.22}$$

where the function has been normalized with the factor $1/2T$ so that

$$\int_{-\infty}^{+\infty} F(t, T)\, dt = 1$$

The Fourier transform of this is

$$\tilde{F}(f, T) = \frac{\sin(2\pi f T)}{2\pi f T} = \mathcal{F}(2f, T) \tag{A5.23}$$

where the function $\mathscr{F}(f, T)$ will be used for function $(\sin \pi fT)/\pi fT$. This function is called the Fraunhofer (diffraction) function because of its relation to the phenomenon of the diffraction of waves in passage through a slit (equivalent to a "square pulse" of transmission in space). Like the Lorentzian, the quantity $2T\mathscr{F}(2f, T)$ has properties that approach the δ-function as $T \to \infty$. It has a height of $2T$, and a HWHM of $0.3017\ldots/T$,

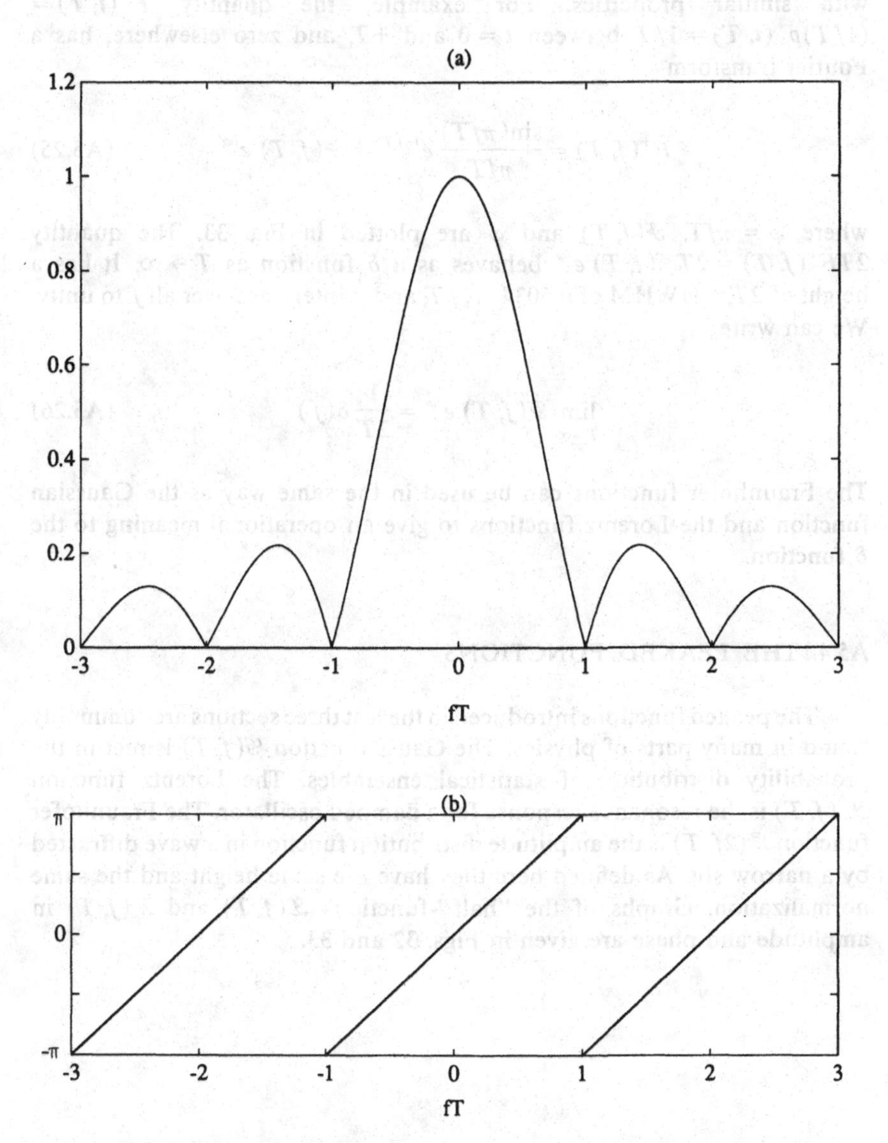

FIGURE 33. (a) The Fraunhofer function $\mathscr{F}(f, T)$ and (b) its phase.

and the integral over all f is unity. Consequently we can write

$$\lim_{T\to\infty} \mathcal{F}(2f, T) = \frac{1}{2T}\,\delta(f) \tag{A5.24}$$

The "half-functions" $p^{+}(t, T)$ and $p^{-}(t, T)$ also have Fourier transforms with similar properties. For example, the quantity $F^{+}(t, T) = (1/T)p^{+}(t, T) = 1/T$ between $t = 0$ and $+T$, and zero elsewhere, has a Fourier transform

$$\tilde{F}^{+}(f, T) = \frac{\sin(\pi f T)}{\pi f T}\, e^{i\pi f T} = \mathcal{F}(f, T)\, e^{i\varphi} \tag{A5.25}$$

where $\varphi = \pi f T$. $\mathcal{F}(f, T)$ and φ are plotted in Fig. 33. The quantity $2T\tilde{F}^{+}(f, T) = 2T\mathcal{F}(f, T)\, e^{i\varphi}$ behaves as a δ function as $T \to \infty$. It has a height of $2T$, a HWHM of $0.6034\ldots/T$, and it integrates over all f to unity. We can write

$$\lim_{T\to\infty} \mathcal{F}(f, T)\, e^{i\varphi} = \frac{1}{2T}\,\delta(f) \tag{A5.26}$$

The Fraunhofer functions can be used in the same way as the Gaussian function and the Lorentz functions to give an operational meaning to the δ function.

A5.4. THE PEAKED FUNCTIONS

The peaked functions introduced in the last three sections are commonly found in many parts of physics. The Gauss function $\mathcal{G}(f, T)$ is met in the probability distribution of statistical ensembles. The Lorentz function $\mathcal{L}^2(f, T)$ is the resonance response for a damped oscillator. The Fraunhofer function $\mathcal{F}(2f, T)$ is the amplitude distribution function in a wave diffracted by a narrow slit. As defined here they have the same height and the same normalization. Graphs of the "half"-functions $\mathcal{L}(f, T)$ and $\mathcal{F}(f, T)$ in amplitude and phase are given in Figs. 32 and 33.

APPENDIX 6

WAVE FUNCTIONS AND BRA-KETS

In Section 7.10 the angular wave functions (spherical harmonics) $Y_l^m(\theta, \phi)$ and the radial wave functions $R_{nl}(r)$ were introduced. We set out here their functional forms.

The normalized spherical harmonics for $m \geq 0$ are given by the general formula

$$Y_l^m(\theta, \phi) = (-1)^m \left[\frac{2l+1}{4\pi} \frac{(l-m)!}{(l+m)!}\right]^{1/2} P_l^m(\cos\theta)\, e^{im\phi} \qquad \text{(A6.1)}$$

where $P_l^m(\cos\theta)$ is the Legendre polynomial. The spherical harmonics of negative m are given by

$$Y_l^{-m}(\theta, \phi) = (-1)^m Y_l^{m*}(\theta, \phi) \qquad \text{(A6.2)}$$

Explicit forms of the spherical harmonics for the lower orders are

$$Y_0^0(\theta, \phi) = \frac{1}{(4\pi)^{1/2}}$$

$$Y_1^{-1}(\theta, \phi) = +\left(\frac{3}{8\pi}\right)^{1/2} \sin\theta\, e^{-i\phi}$$

$$Y_1^0(\theta, \phi) = \left(\frac{3}{4\pi}\right)^{1/2} \cos\theta$$

$$Y_1^1(\theta, \phi) = -\left(\frac{3}{8\pi}\right)^{1/2} \sin\theta\, e^{+i\phi}$$

$$Y_2^{-2}(\theta, \phi) = \left(\frac{15}{32\pi}\right)^{1/2} \sin^2\theta\, e^{-i2\phi} \qquad \text{(A6.3)}$$

$$Y_2^{-1}(\theta, \phi) = \left(\frac{5}{8\pi}\right)^{1/2} \sin\theta \cos\theta \, e^{-i\phi}$$

$$Y_2^{0}(\theta, \phi) = \left(\frac{5}{16\pi}\right)^{1/2} (3\cos^2\theta - 1)$$

$$Y_2^{1}(\theta, \phi) = -\left(\frac{5}{8\pi}\right)^{1/2} \sin\theta \cos\theta \, e^{+i\phi}$$

$$Y_2^{2}(\theta, \phi) = \left(\frac{15}{32\pi}\right)^{1/2} \sin^2\theta \, e^{+i2\phi} \ldots \text{etc.}$$

$R_{nl}(r)$ is the radial wave function. It can be expressed analytically for the case of the hydrogenic atom where the potential energy term in the atomic Hamiltonian is explicitly known:

$$V(\mathbf{r}) = -\frac{1}{4\pi\varepsilon_0}\frac{Ze^2}{r}$$

For this case the radial wave functions have the following explicit forms (the hydrogenic wave functions):

$$\begin{aligned}
R_{10}(r) &= 2\left(\frac{Z}{a}\right)^{3/2} \exp\left(-\frac{Zr}{a}\right) \\
R_{20}(r) &= 2\left(\frac{Z}{2a}\right)^{3/2} \left(1 - \frac{Zr}{2a}\right) \exp\left(-\frac{Zr}{2a}\right) \\
R_{21}(r) &= \frac{2}{\sqrt{3}}\left(\frac{Z}{2a}\right)^{3/2} \left(\frac{Zr}{2a}\right) \exp\left(-\frac{Zr}{2a}\right) \\
R_{30}(r) &= 2\left(\frac{Z}{3a}\right)^{3/2} \left[1 - \frac{2Zr}{3a} + \frac{2}{3}\left(\frac{Zr}{3a}\right)^2\right] \exp\left(-\frac{Zr}{3a}\right) \\
R_{31}(r) &= \frac{4\sqrt{2}}{3}\left(\frac{Z}{3a}\right)^{3/2} \left(\frac{Zr}{3a}\right)\left(1 - \frac{Zr}{6a}\right) \exp\left(-\frac{Zr}{3a}\right) \\
R_{32}(r) &= \frac{2\sqrt{2}}{3\sqrt{5}}\left(\frac{Z}{3a}\right)^{3/2} \left(\frac{Zr}{3a}\right)^2 \exp\left(-\frac{Zr}{3a}\right) \ldots \text{etc.}
\end{aligned} \tag{A6.4}$$

where a has been written for the Bohr radius.

These wave functions have been used in Section 8.3 to evaluate the matrix element $\langle J_iM_i|\hat{\mathbf{r}}|J_kM_k\rangle$. Strictly speaking, however, these wave functions apply to the spatial parts of the system only, whereas the matrix element and the state vectors used to construct it include the spin parts of the wave function. We have really evaluated $\langle L_iM_i|\hat{\mathbf{r}}|L_kM_k\rangle$, where the M here are the quantum numbers of the z component of $\mathbf{L}$. This is justified by the fact that the operator $\hat{\mathbf{r}}$ acts only on the spatial parts of the wave function. The matrix element $\langle\alpha_iL_iM_i|\hat{\mathbf{r}}|\alpha_kL_kM_k\rangle$ can be reduced by the methods discussed in Chapter 7:

$$\langle\alpha_iL_iM_i|\hat{\mathbf{r}}|\alpha_kL_kM_k\rangle = (-)^{L_i-M_i}\begin{pmatrix} L_k & L_i & 1 \\ M_k & -M_i & q \end{pmatrix}\langle\alpha_i \quad L_i\|\hat{\mathbf{r}}\|\alpha_k \quad L_k\rangle \quad \text{(A6.5)}$$

Evaluated for the $k = 2p \rightarrow i = 1s$, $\Delta M = 0$, $q = 0$, π transition in hydrogen, shown in Fig. 34a, this becomes

$$\begin{aligned}
\langle n_iL_iM_i|\hat{\mathbf{r}}|n_kL_kM_k\rangle &= \langle 1 \quad 0 \quad 0|\hat{\mathbf{r}}|2 \quad 1 \quad 0\rangle \\
&= \begin{pmatrix} 1 & 0 & 1 \\ 0 & 0 & 0 \end{pmatrix}\langle 1 \quad L=0\|\hat{r}\|2 \quad L=1\rangle \\
&= -\left(\frac{2.1.1}{1.2.3}\right)^{1/2}\langle 1s\|\hat{r}\|2p\rangle \\
&= -\left(\frac{1}{3}\right)^{1/2}\langle 1s\|\hat{r}\|2p\rangle \qquad \text{(A6.6)}
\end{aligned}$$

The same result, but with a positive sign, is obtained for the σ transitions $\Delta M = \pm 1$, $q = \pm 1$. The reduced matrix element can be evaluated as the radial part of the wave functions

$$\begin{aligned}
\langle 1s\|\hat{r}\|2p\rangle &= \int_0^\infty R_{1s}rR_{2p}r^2\,\mathrm{d}r \\
&= \int_0^\infty 2\left(\frac{Z}{a}\right)^{3/2}\exp\left(-\frac{Zr}{a}\right)\frac{2}{\sqrt{3}}\left(\frac{Z}{2a}\right)^{3/2}\left(\frac{Zr}{2a}\right)\exp\left(-\frac{Zr}{2a}\right)r^3\,\mathrm{d}r \\
&= \sqrt{3}\,\frac{2^8}{3^5\sqrt{2}}\,\frac{a}{Z} \qquad \text{(A6.7)}
\end{aligned}$$

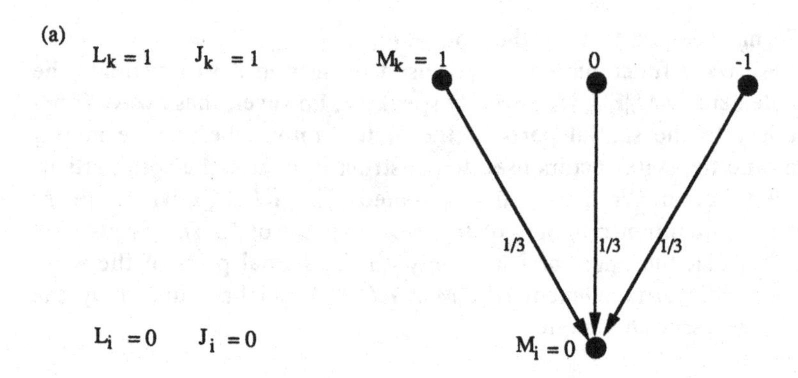

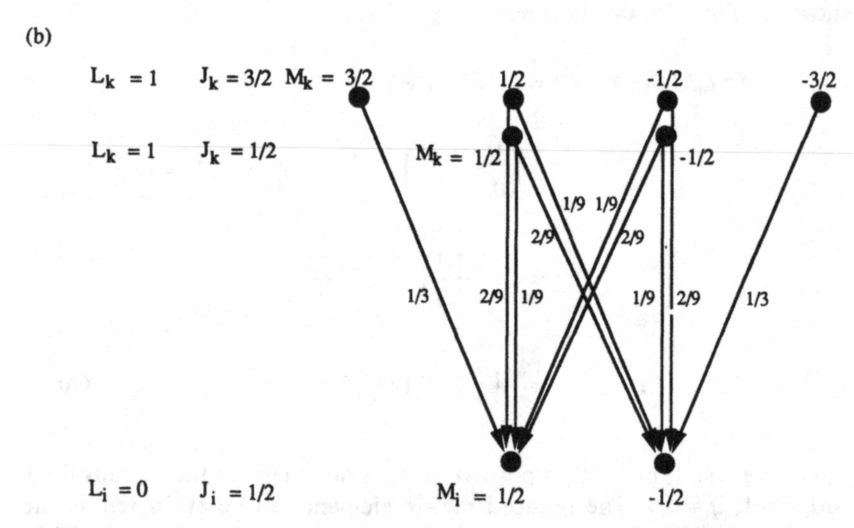

FIGURE 34. Transitions between a p-state ($L_k = 1$) and an s-state ($L_i = 0$). (a) For $S = 0$, i.e., $J = L$. (b) For the transition $2p \rightarrow 2s$ in hydrogen, $S = 1/2$. $L_k = 1, J_k = 3/2$ and $1/2 \rightarrow L_i = 0$, $J_i = 1/2$.

Placed in Eq. (A6.6) this gives the same result (using $Z = 1$ for hydrogen) as Eq. (8.14).

It is useful to demonstrate that the full matrix element

$$\langle \alpha_i \; L_i \; S_i \; J_i \; M_i | \hat{\mathbf{r}} | \alpha_k \; L_k \; S_k \; J_k \; M_k \rangle$$

reduces to the same thing (for this transition $S_i = S_k = S$).

$$\langle\alpha_i \quad L_i \quad S \quad J_i \quad M_i|\hat{\mathbf{r}}|\alpha_k \quad L_k \quad S \quad J_k \quad M_k\rangle$$

$$= (-)^{J_i - M_i}\begin{pmatrix} J_k & J_i & 1 \\ M_k & -M_i & q \end{pmatrix}\langle\alpha_i \quad L_i \quad S \quad J_i\|\hat{r}\|\alpha_k \quad L_k \quad S \quad J_k\rangle$$

$$= (-)^{J_i - M_i}\begin{pmatrix} J_k & J_i & 1 \\ M_k & -M_i & q \end{pmatrix}(-)^{L_i+S+J_k+1}[(2J_i+1)(2J_k+1)]^{1/2}$$

$$\times\begin{Bmatrix} L_i & J_i & S \\ J_k & L_k & 1 \end{Bmatrix}\langle\alpha_i \quad L_i\|\hat{r}\|\alpha_k \quad L_k\rangle$$

The level structure is now more complicated as shown in Fig. 34b. There are six states of the excited level. Two of them, the $|\alpha_k \; J_k \; M_k\rangle = |2\,\frac{3}{2} \pm \frac{3}{2}\rangle$ states, decay by single transitions to the $|\alpha_i \; J_i \; M_i\rangle = |1\,\frac{1}{2} \pm \frac{1}{2}\rangle$ states, respectively. For the transition $n = 2$, $L_k = 1$, $J_k = \frac{3}{2}$, $M_k = +\frac{3}{2} \rightarrow n = 1$, $L_i = 0$, $J_i = \frac{1}{2}$, $M_i = \frac{1}{2}$, for which $q = -1$, the matrix element reduces to $-3^{-1/2}\langle 1s\|\hat{r}\|2p\rangle$, identical with the result obtained when spin was ignored. But the others each have two modes of decay; for example, the excited state $n = 2$, $L_k = 1$, $J_k = \frac{1}{2}$, $M_k = +\frac{1}{2}$ decays to both $n = 1$, $L_i = 0$, $J_i = \frac{1}{2}$, $M_i = +\frac{1}{2}$ ($q = 0$), and to $n = 1$, $L_i = 0$, $J_i = \frac{1}{2}$, $M_i = -\frac{1}{2}$ ($q = -1$). The first of these reduces to $(\sqrt{2}/3)\langle 1s\|\hat{r}\|2p\rangle$, and the second to $\frac{1}{3}\langle 1s\|\hat{r}\|2p\rangle$. At first sight it may seem that these are rather different values; but it must be remembered that the two paths of decay are incoherent with each other, and therefore it is the *intensities* rather than the amplitudes that must be added. We must add the squares as indeed is indicated in Eq. (8.11): $(\sqrt{2}/3)^2 + (1/3)^2 = (1/\sqrt{3})^2$ demonstrating that the decay rate of the state by all routes is the same as the others. In Fig. 34 the transitions are labeled with the squares of the above matrix elements, giving the relative probabilities for the transitions.

All of our reductions have finished with the reduced matrix element $\langle 1s\|\hat{r}\|2p\rangle$ which, as stated above in Eq. (A6.7), can be evaluated if the wave functions are known.

APPENDIX 7

THE DENSITY OF WAVES IN A BOX

We must find expressions for the number of waves that can exist in a volume of space. The result will be deduced for an unpolarized wave of unspecified character and will then be modified for the two cases of interest: (i) the density of modes of an electromagnetic wave, (ii) the density of free-particle states.

Provided that the dimensions of the volume are very large compared with the wavelength, the shape is unimportant; we shall consider a cubical volume $L \times L \times L$. We assume either (a) that the waves form a standing wave pattern inside the box such as would be the case if the waves were electromagnetic and the walls of the box were conducting, or (b) that the wave pattern is identical (apart from a possible inversion of phase) in every volume L^3. In both cases the set of wave fronts separated by $\lambda/2$ (nodal fronts in the case of standing waves) must form an integral number of intercepts along each of the sides of length L. Consider a wave (of wavelength λ) whose propagation vector $\mathbf{k}^0$ has direction-cosines $\cos\alpha$, $\cos\beta$, $\cos\gamma$ with respect to the x, y, z axes, respectively. The wave fronts make an intercept of $\lambda/(2\cos\alpha)$ on the x axis, and similarly on the others. The integer numbers of intercepts along the axes are given by

$$n_x = \frac{2L}{\lambda}\cos\alpha, \qquad n_y = \frac{2L}{\lambda}\cos\beta, \qquad n_z = \frac{2L}{\lambda}\cos\gamma \tag{A7.1}$$

From the property of the direction cosines we have

$$n_x^2 + n_y^2 + n_z^2 = \frac{4L^2}{\lambda^2} \tag{A7.2}$$

This is a three-dimensional extension of the simple one-dimensional result $n_x = 2L/\lambda$ appropriate to the case of a standing wave on a string. The number n_x is the number of sectors of vibration along the string—or the harmonic number.

The integers n_x, n_y, n_z define the possible directions of propagation of the waves in the space provided for them:

$$\cos^2 \alpha = \frac{n_x^2}{n_x^2 + n_y^2 + n_z^2}, \qquad \text{etc.} \tag{A7.3}$$

For a box where $L \gg \lambda$, at least one of n_x, n_y, n_z must be very much greater than unity. The set of integers n_x, n_y, n_z define the *mode* of one of the waves in the box.

We wish now to establish an expression for the number of modes that can exist in the box with wavelengths between λ and $\lambda + d\lambda$. The number with wavelengths greater than λ (i.e., with wave numbers less than $1/\lambda$) can be established by the following argument. The set of all possible values of n_x, n_y, n_z can be represented by points at the corner of a cubic lattice of cells of unit length. In this space the number of sets of values of n_x, n_y, n_z required is just the number of such points that lie within the octant of a sphere of radius $2L/\lambda$. This number is

$$\frac{1}{8}\left(\frac{4\pi}{3}\frac{8L^3}{\lambda^3}\right) = \frac{4\pi L^3}{3\lambda^3}$$

and the number in the range λ to $\lambda + d\lambda$ is

$$L^3 n(\lambda)\, d\lambda = L^3 \frac{4\pi}{\lambda^4}\, d\lambda$$

These waves are traveling isotropically in all directions, so that we can generalize the result to the number per unit solid angle by multiplying by $d\Omega/4\pi$. The number of modes of the wave within a volume L^3 of space, within a solid angle $d\Omega$, and within the wavelength range from λ to $\lambda + d\lambda$ is

$$\frac{dn(\lambda)}{d\Omega} L^3\, d\lambda\, d\Omega = \frac{1}{\lambda^4} L^3\, d\lambda\, d\Omega \tag{A7.4}$$

A7.1. THE DENSITY OF ELECTROMAGNETIC FIELD MODES

Electromagnetic waves have transverse polarization and therefore the above result must be multiplied by 2 to allow for the independent polarization states. Equation (A7.4) then can be rewritten as

$$\frac{dn(\lambda)}{d\Omega}\, d\lambda = \frac{2}{\lambda^4}\, d\lambda \tag{A7.5}$$

or, using $\nu = c/\lambda$, as

$$\frac{dn(\nu)}{d\Omega}\, d\nu = \frac{2\nu^2}{c^3}\, d\nu \tag{A7.6}$$

If the distribution is isotropic, these can be written as

$$\begin{aligned} n(\lambda)\, d\lambda &= \frac{8\pi}{\lambda^4}\, d\lambda \\ n(\nu)\, d\nu &= \frac{8\pi\nu^2}{c^3}\, d\nu \end{aligned} \tag{A7.7}$$

A7.2. THE DENSITY OF FREE PARTICLE STATES

We must modify the results above so that they are expressed in terms of the kinetic energy K of the free particle. The wavelength is related to the momentum p of the particle by

$$\frac{1}{\lambda} = \frac{p}{h}$$

$$= \left(\frac{K^2 + 2Kmc^2}{c^2h^2}\right)^{1/2} \quad \text{using the relationship between kinetic energy and momentum,}$$

$$= \left(\frac{2Km}{h^2}\right)^{1/2} \quad \text{using the nonrelativistic limit.} \tag{A7.8}$$

Therefore we have

$$\frac{d\lambda}{\lambda^4} = \frac{\sqrt{2}\, m^{3/2} K^{1/2}}{h^3}\, dK \tag{A7.9}$$

Substituting this in Eq. (A7.4) we obtain

$$\frac{dn(K)}{d\Omega}\, dK = \frac{dn(\lambda)}{d\Omega}\, d\lambda = \frac{d\lambda}{\lambda^4} = \frac{\sqrt{2}\, m^{3/2} K^{1/2}}{h^3}\, dK \tag{A7.10}$$

For an electron that has achieved the free state by ionization from an atom, this may be expressed in terms of the frequency ν of the state using $K = h(\nu - \nu_I)$ where $h\nu_I$ is the energy of the ionizing level with respect to the ground state:

$$\frac{dn(\nu)}{d\Omega}\, d\nu = \frac{dn(K)}{d\Omega}\, dK = \sqrt{2}\left(\frac{m}{h}\right)^{3/2} (\nu - \nu_I)^{1/2}\, d\nu \tag{A7.11}$$

[illegible] of Wave [illegible] A box [illegible]

or, using [illegible]

$$\frac{dn}{dE} = \frac{2\pi}{\cdots} [illegible] \qquad (A.7.6)$$

The distribution [illegible] written as

[illegible] (A.7.7)

[illegible] OF [illegible]

We now relate the results above to [illegible] expressed in terms of the kinetic energy E of the [illegible] related to the momentum p of the particle by

[illegible]

[illegible] the relationship between kinetic energy [illegible]

[illegible] (A.7.8)

Then [illegible] we [illegible]

[illegible] (A.7.9)

Substituting this in Eq. (A.7.6) we obtain

[illegible] (A.7.10)

[illegible] may be expressed in terms of [illegible] the state [illegible] with respect to the ground state:

[illegible] (A.7.11)

Index

Absorber theory, 201
ac Stark effect, 138
Action at a distance, 2
Angular momentum, 53, 223–224
  coupling, 70
Antibunching, 176, 179
Aspect experiments, 191
Atomic states,
  allowed, 59
  angular momentum, 60
  configuration, 68
  expectation value, 85
  operators, 73
  singlet, 72
  state vectors, 62–64
  stationary, 59
  triplet, 72
  vector labels, 62
Autocorrelation, 36

Bohr magneton, 67
Bohr radius, 62
Bohr theory, 61
Boltzmann distribution, 90
Bra-vector, 63
Bunching, 176

Cascade transitions, 191–199
Casimir effect, 217
Chaotic light field, 171, 173
Coherence, 161
  function, first order, 163, 173
  function, second order, 164, 173
  partial, 167
  time, 169, 171
Commuting operators, 66
Compton
  effect, 45–56
  energy and momentum changes, 52–54
  period of the electron, 46
  wavelength of the electron, 46
Configuration, 67–73
Correspondence principle, 83
Counting statistics, 161–180
Cross-correlation function, 36

Damping (decay) constant, 92–95, 105, 110, 116, 118–120
  classical, 25, 105
Delta function, 32, 225–227
Density
  of field modes, 242
  of free-particle states, 243
  of waves in a box, 241–243
Doppler effect, 169
Dynamic Stark effect, 138

Eigenstates, 64
Eigenvalues, 64
Einstein *A* coefficient, 26, 89–95
  classical theory, 91–92
  quantum theory, 105–120
Einstein *B* coefficient, 89–95
  classical theory, 95
  quantum theory, 102–104
Einstein–Podolsky–Rosen, 188
Electric field vector, 2
  of accelerating charge, 7
Electromagnetic theory, 2
Electron classical radius, 62
Energy density of radiation, 35–43
  broad-band field, 206, 214
  modulated field, 208

Energy density of radiation (*Cont.*)
monochromatic field, 207
zero-point field, 219
Ensemble averaging, 211–215
Exponential
function, 228
time averaging, 205

f value, 94
Fine structure constant, 62
Fluorescence, 142
Fourier transformation, 20
Fraunhofer factor, 204
Fraunhofer function, 232
Frequency-propagation space, 212

Gaussian distribution function, 169, 225

Hamiltonian operator, 64, 66, 73, 79–83
atomic, 64, 108
decay, 106
radiation reaction, 108
zero-point, 111
Hydrogen, photoelectric cross section, 156

Intensity
interferometer, 164
of radiation, 35–43
monochromatic, 30
polychromatic, 33
spectral, 33
Interference of light, 161
Ionization energy, 142
Irradiance, 35

Ket-vectors, 63

Lamb shift, 116
Landé factor, 67
Larmor's formula, 11, 23
Lifetime, 93
Light
classical and nonclassical, 176–179
emission and detection, 184–188
nature of, 1–5, 181–201
velocity of, 4
Lorentz factor, 205, 229
Lorentz resonance function, 28, 228–231
Lorentz transformations, 48, 49, 50, 219, 220
Lorentz invariance, 110, 218
of zero-point field, 219
Luminal frame, 188–189

Magnetic field vector, 2
of accelerating charge, 8
Matrix elements
of atomic operators, 73–76
of decay Hamiltonian, 120
Maxwell's equations, 3
Momentum of recoil, 52, 56, 112
Multiplicity, 70
Mutual coherence function, 36, 212

Operators, 73
angular momentum, 66
expectation value, 85
Hamiltonian, 64, 79–83
momentum, 85
type $\hat{T}$, 74
Optical transition, 121–140
Optical coherence, 161–180
Optical pumping, 121
Orbital
angular momentum, 60, 61, 69, 70
radius, 61
Oscillator
free, 13
polychromatic, 19–21
Q value of, 30
radiation from 25–26
steady-state forced, 14

Parity, 77
Pauli exclusion principle, 68
Photoelectric
antibunching, 176, 179
bunching, 176
cross section, 145–156,
for hydrogen, 156–159
effect, 141–160
statistics, 174–176
Photoconductivity, 142
Photon, 144
Planck's constant, 54
Planck distribution, 91
Poisson distribution, 169, 174
Polarization
defining parameter, 18
vectors, 14, 17–19
Poynting vector, 11, 35–43
Probability amplitude, 80

Q value of oscillator, 30
Quantum rules for interaction, 54
Quantum structure of atom, 57–87

Rabi oscillations, 99, 122–126
Radial wave function, 236
Radiated power, 25
from a driven oscillator, 30, 32
Radiation
from driven atom, 136–140
from driven oscillator, 26
of energy, 11, 25–26
field, 7–8
reaction force, 22
Radiative lifetime, 92–95
Rayleigh scattering cross section, 31
Relativity, theory of, 4
Resonance functions
$\mathscr{L}$, 28, 228–231
$\mathscr{R}$, 28–29
Retarded time and position, 7–10
Ritz combination principle, 58
Rotating wave approximate, 99, 123
Rydberg constant, 61

Scattering, 27–33
of broad-band radiation, 31
cross section, 30
of monochromatic radiation, 27–31
of radiation by a free charge, 45–56
Schrödinger equation, 79
Selection rules, 58
electric dipole, 76–79
Spectral energy density, 41–43, 207
black-body radiation, 215
zero-point field, 215, 219
Spectral intensity, 33, 35–43, 207
Spectral lines, 58
Spherical harmonics, 86, 235
Spin angular momentum, 60, 69, 70
Spontaneous decay, 105–120
constant, 105, 110, 116
Stark effect, 67
dynamic or ac, 138
State symbols, 70
State vectors, 62–64
equation of motion, 79
labels, 65–73
Stationary states, 59
time-dependent form, 80
Statistical weight, 89
Stimulated absorption and emission, 97–104
Superposition states, 79

$\hat{\mathrm{T}}$-type operators, 74
Thomson scattering cross section, 31
Time averaging, 203–210
exponential, 205
truncated, 204
Transitions in atoms, 57–59, 121–140
electric dipole, 76–79
photoelectric, 59, 141–180
semi-classical theory, 97–104
spontaneous emission, 59–90
stimulated absorption and emission, 58, 90
Truncated time averaging, 204
Two-level atom, 121
with decay, 126–131
probability of excitation, 131
steady-state solution, 131–135

Wave function, 84–86
atomic states, 86
free particle, 85
for hydrogen, 235–239
radial, 86, 236
spherical harmonics, 86, 235
Wigner 3-j and 6-j symbols, 75, 76
Work function, 142

Zeeman effect, 66
Zero-point vacuum field, 110, 217–222
Hamiltonian, 111